The ESP Enigma

The Scientific Case for Psychic Phenomena

Diane Hennacy Powell, M.D.

Walker & Company
New York

Published by Walker Publishing Company, Inc., New York

ART CREDITS

1. Used by permission of Russell Dewey, Ph.D. 2. Diagram by Diane Hennacy Powell, M.D. 3. Diagram by Diane Hennacy Powell, M.D. 4. Reprinted with permission by ACSM/CaGIS from "Interrupting the World" by Richard E. Dahlbert in *Matching the Map Projection to the Need*, figure 2-4. 5. Photograph by Daniel Fennel, M.D. 6. Photograph by Daniel Fennel, M.D. 7. Andrew McClenaghan/Science Photo Library.

All papers used by Walker & Company are natural, recyclable products made from wood grown in well-managed forests. The manufacturing processes conform to the environmental regulations of the country of origin.

LIBRARY OF CONGRESS CATALOGING-IN-PUBLICATION DATA

Powell, Diane Hennacy.
The ESP enigma : the scientific case for psychic phenomena / Diane Hennacy Powell.
p. cm.
Includes bibliographical references and index.
ISBN-13: 978-0-8027-1606-4 (hardcover)
1. Consciousness. 2. Parapsychology. I. Title.
BF311.P67 2009
133.8—dc22
2008018533

Visit Walker & Company's Web site at www.walkerbooks.com

First published by Walker & Company in 2009
This paperback edition published in 2010

Paperback ISBN: 978-0-8027-1028-4

1 3 5 7 9 10 8 6 4 2

Typeset by Westchester Book Group
Printed in the United States of America by Quebecor World Fairfield

Praise for *The ESP Enigma*

"*The ESP Enigma* is an unflinching examination of the puzzle of consciousness—how it's able to do things that, according to our textbooks, don't make sense. Refreshingly clear, scientifically accurate, up to date, and comprehensive, this book shatters conventional beliefs about the nature of the mind and reality itself."

—Dean Radin, Ph.D., senior scientist, Institute of Noetic Sciences, and author of *The Conscious Universe* and *Entangled Minds*

"A cogent argument offering many striking examples of the power and potential of the unconscious . . . Powell's theory of consciousness seeks to resolve some of these mysteries." ***—Kirkus Reviews***

"A tour de force that is more comprehensive and persuasive than anything I have read in years."

—Stanley Krippner, Ph.D., coauthor of *Extraordinary Dreams and How to Work with Them*

"Wonderfully profound, deeply accessible, *The ESP Enigma* carefully unpacks the folds and fields of extended human consciousness to reveal the sometimes subtle, often surprising, and ever fascinating mental dimensions that make us what we are—yet which (for now) Science still doggedly refuses to accept."

—Maj. Paul H. Smith, U.S. Army (ret.), president of the International Remote Viewing Association and author of *Reading the Enemy's Mind: Inside Star Gate, America's Psychic Espionage Program*

"Not about to compromise her credibility by claiming that mind power can levitate tables, Powell adheres to the possibility that human consciousness might affect matter at the quantum mechanical level. Incorporating Powell's knowledge of neuroscience, this work should appeal to those open to the idea that ESP exists."

—Booklist

"The wonderful syntheses of the declassified information from the Cold War era, case reports, and the burgeoning information arising from physiological studies with new brain-scanning techniques makes for an intriguing discussion of the process of consciousness . . . Few authors in the field of consciousness research have the depth and breadth to present the larger view that Dr. Hennacy Powell offers with eloquence."

—Dan F. Fennell, *Journal of Humanistic Psychology*

"Finally, scientists and laymen alike have a clear and compelling introduction to the controversial world of contemporary psychic and consciousness research. Dr. Powell's *The ESP Enigma* is destined to become a classic in the field."

—Gary E. Schwartz, Ph.D., director, Laboratory for Advances in Consciousness and Health at the University of Arizona, and author of *The Energy Healing Experiments*

"*The ESP Enigma* will help open people's minds not only to accept what they can't explain but also to learn what is already known about ESP. If we spent more time and money exploring our inner world, we would better understand the mysteries of life."

—Bernie Siegel, M.D., author of *Love, Medicine & Miracles* and *365 Prescriptions for the Soul*

To my daughter Allie,
whose presence lights up the room
and in loving memory of Nikki,
my canine companion and muse

CONTENTS

INTRODUCTION

WHETHER WE CONSIDER ourselves believers in psychic phenomena or not, many of us have had something happen to make us wonder about the subject. It could have been someone telling us that she was just thinking about us when we called, or vice versa. It might have been a gut feeling to drive a different route from our usual one, only to discover later that a large accident occurred on the road we didn't take. Such experiences may not happen often, but they can leave us with a profound feeling that we are interconnected, that we can know things without understanding how, and that there must be more to our universe than we detect through ordinary senses.

People have believed in psychic abilities since the beginning of recorded history. Certain individuals report more experiences with psychic phenomena than others. Since these experiences usually only occur spontaneously for most of us, many cultures developed divination aids in order to access psychic information more readily. The Dogon in West Africa toss cowrie shells into a basket and interpret the patterns. The Chinese devised the I Ching, and Egyptian priests slept in special temples in order to have prophetic dreams.

Perhaps the most famous divination practice was the Delphic

oracle, who drew the rich and famous from all over the Greek world from the sixth century B.C. until the fourth century A.D. The Greek historian Herodotus claimed that the Delphic psychic spoke in a trance induced by natural gases that seeped through the rocks. This was discounted as a myth until 2001, when Jelle de Boer, a geologist at Wesleyan University in Middletown, Connecticut, analyzed the hydrocarbon gases emitted by the temple's nearby spring. He reported in *Geology* that he found ethylene in sufficient concentrations to have created a narcotic effect that would have been experienced as a floating or disembodied euphoric state.

The most widespread form of divination is scrying, from the old word *descry*, which means "to catch sight of" and involves deep concentration on a smooth reflective surface until an image appears. Ancient Greeks looked for answers in spring waters; in ancient India, warriors peered into vessels filled with water to see if they'd return from battle; Tahitians poured water into a hole at crime scenes to scry the image of the culprits. The most famous tool for scrying has been the crystal ball, which became a tool of Gypsies, among others.

The Old Testament of the Judeo-Christian Bible contains numerous accounts of prophets, but Christianity forbade all forms of prophecy except for divine revelation and astrology. As Christianity spread, many forms of prophecy declined or went underground in Christian areas, lest the practitioners be accused of heresy or witchcraft. In the Middle Ages, popes still consulted astrologers to provide them with propitious dates for coronation, but after the Copernican revolution changed our understanding of planetary movements, the Catholic Church declared divine revelation to be the only acceptable form of prophecy.

Westerners' growing disbelief in psychic abilities was influenced by the development of the scientific method. During the eighteenth-century Age of Enlightenment, the universe became increasingly viewed as a mechanistic system, accurately known only through observation, calculation, and reason. Anything associated with the supernatural or psychic phenomena lost credibility.

Skepticism about psychic phenomena was further fueled by scandals that linked claims of psychic abilities with con artists who preyed upon people's vulnerabilities. Also, as the psychiatric profession arose, reports of psychic experiences were often accompanied by signs of irrational thinking and became interpreted as signs of brain pathology, rather than innate gifts or capacities.

Added to this was the belief that the mind exists solely within the brain. This is an idea that has grown since François de La Peyronie, an eighteenth-century French surgeon, observed changes in human behavior that accompanied specific brain injuries. The scientific model of the brain and consciousness that evolved in this historical context did not have to account for psychic phenomena.

The scientific model is based on these facts: The brain is a biological machine with over a hundred billion neurons, or brain cells, each of which has an average of five thousand connections to other neurons. Electrical signals pass along the neurons, causing them to release chemical messengers, such as serotonin and dopamine, from their terminal ends. These messengers land on the receptors of neurons on the other side of the synapse, or region between neurons for chemical connection. Once neurons receive enough stimulation from their connecting neurons, they send signals along their axons to other neurons. There are almost an infinite number of possible patterns of activity along the neuronal network, and specific patterns are believed to represent concepts, thoughts, or memories. Francis Crick, the late codiscoverer of

DNA's structure, summarized this model when he said, "The astonishing hypothesis is that 'You,' your joys and your sorrows, your memories and your ambitions, your sense of personal identity and free will, are in fact no more than the behavior of a vast assembly of nerve cells and their associated molecules."[1]

Even though scientists, including Crick, admit that they do not know what consciousness is or how it is generated, proponents of the current model consider consciousness to be a byproduct of a brain that can access new information only by direct sensory input. The body has receptors for sound, taste, sight, touch, smell, and proprioception (detection of body movement and placement), but there is no hardware to access sensory information from distant points in space and time, let alone to send information directly from one brain to another. The current concept of consciousness cannot accommodate the existence of psychic abilities, and as rational beings, we are skeptical of that which cannot be explained scientifically.

Yet some psychic phenomena have been measured and verified scientifically. One example is the work by Adrian Parker and Joakim Westerlund at the University of Gothenburg in Sweden. They placed the "receivers" of telepathic information in isolation and minimized their sensory input, thereby preventing any potential interference. The "senders" sat in an isolated room watching a film, while the receivers simultaneously commented upon what information came to mind. A real-time recording of the receivers' comments was then superimposed upon the transmitted film for analysis. One participant described accurately, in real time, a full sequence of events as they occurred in the film.[2]

Another example is the research at Stanford Research Institute by Russell Targ and Hal Puthoff, two laser physicists, which provided valuable information to almost every branch of the U.S.

intelligence community during the Cold War with the Soviet Union. Much of their work was done on remote viewing, in which the sender went to an undisclosed location and the receiver drew a picture of it. One of their best receivers was Pat Price, a retired policeman who had helped the Berkeley police in their search for Patty Hearst. In his first attempt at remote viewing for SRI he achieved 90 percent accuracy in his psychic drawing of a swimming pool complex that included its dimensions, size, location, and the function of the pools and adjacent buildings.

Despite such experiments, the scientific community still questions the validity of psychic phenomena, demanding research data that is reproducible under tightly controlled conditions in order to accept phenomena as true. At least on a public level, most scientists have taken the stand that something as extraordinary as psychic phenomena requires the data to be extraordinary as well.

A critical review of the laboratory data for psychic phenomena reveals cumulative data would have been sufficient evidence for other areas of research. If one wants to prove whether or not telepathy can exist, one strong convincing case for its existence should be sufficient, because that is analogous to one living brontosaurus being proof that the species isn't extinct. William James, the late professor of psychology at Harvard, shared this same view on what is sufficient proof. He described paranormal experiences as "white crows" and said that "if you wish to upset the law that all crows are black, you must not seek to show that no crows are [black]; it is enough if you prove one single crow to be white."[3]

Applying James's analogy to the status of psychic research, there have been several sightings of white birds. Scientists haven't disputed that they are white, just whether they are crows. One has to capture the white bird, inspect it closely, and perhaps even test its DNA to prove that it is a crow. Anything short of this would be insufficient for a scientific revolution. Technology has advanced such that we can better identify the "white bird" in psychic research, and it does appear to be a crow.

But proof of the existence of some psychic phenomena would mean we need to reconcile how they are possible given our understanding of consciousness and the brain. This would pose more of a challenge if the current model was complete and psychic phenomena were the only mystery. Instead, relatively little is known about consciousness. For example, no one has been able to answer what has been called the "hard question" of consciousness: how can something as nonmaterial as consciousness arise from something material like the brain? The model also doesn't explain free will or our feeling that there is an "I" that has experiences. On top of that, there are reports of near-death survivors that suggest that consciousness can continue even when the brain has shut down, whereas the current scientific paradigm continues to regard consciousness as a product of brain chemistry and wiring.

A primary reason psychic phenomena are hotly contested by the scientific community is that the validity of such phenomena would mean a major scientific revolution, similar to the Copernican revolution that forced us to accept the sun as the center of the solar system. Scientific revolutions are not easy matters. Thomas Kuhn, the late physicist and professor of the history of science at MIT, compared scientific revolutions to political revolutions, with good reason. They involve a lot of politics. Some interested scientists have openly stated that they were afraid that they would lose their credibility should they investigate psi, the technical term for psychic abilities. Partly as a result of these concerns, today there are no more than fifty scientists across the globe involved full-time in this area of research. But it is the study of anomalies, such as psychic experiences, that will provide a better understanding of consciousness.

When a scientist has devoted his or her career to studying psychic abilities, it has usually been because of a thought-provoking personal experience. One of many examples is Hans Berger, the inventor of the electroencephalogram (EEG), which is used clinically to measure brain waves. Berger invented this device as a means of investigating telepathy after an extraordinary experience with his sister, who sent him a telegram saying she was very concerned that something bad had happened to him. Her timing was impeccable. Earlier that day he was almost killed while riding a horse. His sister's timely concern was so striking that Berger hypothesized that brains must be capable of sending signals to one another. Because this was during the time when electromagnetism was an exciting new field of inquiry, he thought that he'd find the answer by designing a machine that measures the electromagnetic activity of the brain. Although the EEG did not provide proof of telepathy, it has been of great help in advancing our understanding of the brain.

My own interest dates back to when I was thirteen years old. Through a good friend, I met a circus magician known primarily for his Houdini escapist tricks. In my friend's living room, he demonstrated something astonishing. From twenty feet across the room, the magician read, word for word, the contents of any book that I randomly chose from among hundreds on the bookshelves. There were no mirrors behind me, and I knew that these books belonged to my friend, not the magician. Even if he had memorized all of the books, he would also have needed exceptional luck to guess which pages I chose. There was no rational explanation at the time for what I observed, but it fostered a deep, abiding curiosity.

I was already familiar with extraordinary mental abilities in one sense. I was a math prodigy as a child, someone who could do ninth- and tenth-grade math at seven years of age. And at age four my grandmother was a musical prodigy who could play songs accurately after hearing them only once. Much later I learned of autistic savants and other prodigies whose abilities were well documented but, like psychic phenomena, were not explained by the current understanding of consciousness and the human brain.

My interest led me to study neuroscience in college and specialize in neuropsychiatry at the Johns Hopkins University School of Medicine. While on faculty at Harvard Medical School, I encountered a patient who claimed to be psychic. She then told me several accurate details about my life and made specific predictions about my future, all of which eventually came true. After this encounter, I decided to systematically investigate psychic phenomena. And over the past twenty years I have gained invaluable insight from patients who shared details of their psychic experiences.

The ESP Enigma presents a summary of the research on the four basic psychic abilities: telepathy (the ability to access someone else's consciousness), psychokinesis (the ability for one's conscious intention to directly act upon physical matter), clairvoyance (the ability to see something remote in space or time), and precognition (the ability to access the future). Some studies looked at large groups of individuals with the hypothesis that psychic abilities may be an innate capacity in all of us. Others have researched individuals who seem to possess these abilities to an extraordinary degree.

The book also addresses another question: how could psychic phenomena be possible? There have been enough advances in science over the last twenty years to now propose an acceptable mechanism by which psychic phenomena could occur. This new model for the brain and consciousness has the potential to reshape not just our attitudes toward psychic phenomena but also our understanding of our own minds.

Chapter 1

CONSCIOUSNESS AND THE BRAIN

The highest activities of consciousness have their origins in the physical occurrences of the brain, just as the loveliest melodies are not too sublime to be expressed by notes.

—W. Somerset Maugham

We all experience consciousness, yet it remains one of life's greatest mysteries. People even disagree on the most fundamental questions: What is consciousness? What is it made of? No one can say for sure, other than to define it as the stream of thoughts and feelings we experience while awake. That definition leaves unanswered the primary question of its essence. Our thoughts are as ephemeral as fairy dust. They bubble out of our deep pool of consciousness, but is that pool a form of energy, something material, another force of nature, or something else?

Other hotly debated questions are: Does the brain actually create consciousness? Or does it merely process it, or transmute it into its myriad forms? And how does our inner experience of the world, shaped by our brains, actually relate to external reality? Our sensory organs and brain limit us to a narrow slice of the full spectrum of sounds and sights, but do they limit us in other, unknown ways?

Philosophers and theologians have asked these questions for millennia, initially in the context of paradigms that made psychic phenomena possible. But during the past century, the questions have led neuroscientists to develop a paradigm that deemed psychic phenomena impossible. Yet over the past century there also have been discoveries in subatomic physics that have made possible a conceptual framework for psychic phenomena, even though many scholars of consciousness have not incorporated these discoveries into their thinking.

Questions about consciousness are the foundation for this book, and the theories behind their answers largely determine how easy or hard it is to accept psychic phenomena. When one analyzes the evolution of theories, one sees that facts usually drive theories, but theories often drive what is accepted as a fact. Ideally, theories and facts evolve together, and new models arise that include facts derived under old assumptions while discarding ideas outdated by new information. But people can have difficulty separating assumptions from facts. As you read this book, think about what you really know and how you know it. Some beliefs might turn out to be something that you just assume.

MONISM VERSUS DUALISM

Philosophers divide ways of looking at consciousness into two basic categories: monism and dualism. Almost every academic devoted to the study of consciousness aligns him- or herself into one of these camps.

In monism, one universal and unified set of laws underlies nature. The mental and physical realms cannot be separated because they are one and the same. Mentalism, the most ancient form of monism, regards the mind as the only thing that is real. Mentalism dates back thousands of years to the earliest Eastern philosophers. They studied consciousness by meditative techniques that enabled direct experience of various levels of consciousness. The Hindu and Buddhist belief that everything is pure consciousness results from mystical experiences that are so powerful they feel like they reveal the "real reality." According to this view, our usual perception of the world is illusory and everything is really "one," or inseparable.[1] This perspective doesn't just allow for psychic phenomena; it makes them very likely because there is no difference or separation between one's internal and external reality. Everything is simply a product of our mind, and therefore everything is possible.

In dualism, the mental and physical are separate and radically different from each other. Like the relationship of monism to Eastern religions, dualism has been a component of Western religions. Both Plato and Aristotle were dualists with a large influence on Western religion. Plato was among the first to propose that we have souls trapped in our bodies.

René Descartes is another famous dualist. In 1641 he formulated a mind-brain distinction that drew upon the qualitative differences between conscious experience and physical matter. Consciousness does seem nonmaterial and unlike anything in the physical world. Also, during conscious experience we have the compelling feeling that there is an "I" that exists, witnesses, and acts upon our thoughts. We also feel that there is a separate and external world, and that we have free will to act within it.

Descartes identified the pineal gland as the place in the brain where the soul, or mind, could meet and affect the physical.[2] Descartes' dualism is highly compatible with psychic phenomena and the many reports of psychics that their consciousness leaves their body in order to either access information or influence the physical world. His dualistic theory endured for centuries because it meshed comfortably with many religious perspectives.

But within science, Descartes' dualism had a major deficiency. Descartes' theory was sarcastically referred to as the "ghost in the machine" in 1949 by the British philosopher Gilbert Ryle. In his book *The Concept of Mind*, Ryle argued that dualist systems like Descartes' were absurd because there was no means for the body and mind to interact. To Ryle, Descartes' theory implied that the body was a biological machine magically controlled by a soul, or "ghost."

William James was another prominent dualist. He personally experimented with nitrous oxide, or laughing gas, which induced experiences that felt like his consciousness left his body. These led him to publicly wonder in 1898 whether the brain actually produced consciousness. In the following statement he proposed that the brain could just be a means for transmitting consciousness:

> Just as a prism alters incoming white light to form the characteristic colored spectrum, but is not the source of the light; and just as the lengths of the pipes of an organ determine how the inflowing air yields certain tones and not others, but are not, themselves, the source of the air, so too the brain may serve a permissive, transmissive, or expressive function, rather than solely a productive one, in terms of the thoughts, images, feelings, and other experiences it allows.[3]

Aldous Huxley said almost the same thing in 1954.[4] He was another dualist who considered the brain to be a filter that primarily blocks out consciousness rather than actually generating it. He saw the brain as only allowing us to register and express a narrow range of reality, which could be broadened during states of altered consciousness, such as meditation and dreams.

Around the same time as Descartes, a second form of monism, neutral monism, was proposed by the philosopher Baruch de Spinoza. It regards both the mental and physical as reducible to some third entity, which was called "God" or "nature" during Spinoza's time. Because neutral monism describes the abstract mystery of consciousness in terms of another abstract mystery, it is unsatisfactory to those who want concrete answers. As for psychic phenomena, the theory is neither compatible nor incompatible with them. It doesn't rule them in or out.

The third and most recent monism is the perspective most neuroscientists share. Materialism assumes that only the physical world is real and that the mind can be reduced to or equated with something physical. Materialists agree with W. Somerset Maugham that our brains produce the streams of consciousness that can take us to an imaginary tropical island or vividly recall Thanksgiving at Grandma's house when we were eleven years old. In *The Astonishing Hypothesis,* Francis Crick expressed this belief that our consciousness and sense of self result strictly from chemical and electrical brain processes. Because materialism is not consonant with many nonscientists' beliefs, Crick called it an "astonishing hypothesis." Psychic phenomena are not possible in materialism, because there is no mechanism for consciousness within one brain to communicate remotely with the consciousness of another. Materialism also doesn't allow one to see the future or to influence the physical world solely by conscious intention.

THE ASCENT OF MATERIALISM

Materialism became the primary academic model for consciousness because of compelling evidence that linked the brain to consciousness. Damage to the brain can cause consciousness to be lost. Strokes in the brainstem can render someone comatose. Also, grand mal seizures lead to a temporary loss of consciousness that is always accompanied by specific changes in a person's EEG.

Very discrete injuries to the brain lead to specific alterations of conscious experience. For example, a tiny stroke in the area of the visual cortex that processes color (V4) leads to vision where our world looks like we've entered a black-and-white movie. The neurologist Oliver Sacks wrote about a patient whose damage to V4 caused him to lose interest in sex because his wife's skin appeared to be gray to him, as though she were dead. Food became unappealing because it was gray, a color we naturally associate with decay.

A stroke in the cortical area called MT specifically eliminates the ability to see motion. As a result, any moving object is seen as a series of static images, and the person with an MT stroke cannot gauge how quickly something is moving. The neurologist V. S. Ramachandran described the ways this impacts one's life. A patient with an MT stroke became afraid to cross the street if any cars were driving on it, and her glass overflowed when she poured herself a drink.

Some brain structures are intimately involved with the content of our conscious thoughts. For example, the hippocampi are necessary to consolidate long-term memory. If both hippocampi are damaged, a person becomes frozen in time. Although he is able to remember everything that happened prior to the time of damage, he is unable to form new memories. He can appear normal at first, but if you ask him about something from just a few minutes before, he has no memory of it. He won't tell you that he forgot, because these patients confabulate, or make something up, to answer your questions. They aren't liars. Their brains create information that they actually believe. It is like our brains' automatic filling in of blind spots in our field of vision, rather than letting us see areas of black. When a patient is living in a continuously growing time gap, his brain will confabulate as a way of preserving his sense that his stream of consciousness is continuous.

Other evidence for the brain's connection to consciousness is the replication of specific conscious experiences by direct electrical stimulation of brain areas. During the 1940s and 1950s the Canadian neurosurgeon Wilder Penfield mapped out large sections of the brain by applying direct electrical stimulation to brain cells in the operating room before he removed diseased brain tissue. Because the brain doesn't have pain receptors, he could keep patients awake and ask them about their experiences during the stimulation. Among other findings, Penfield outlined the regions called the homunculi, or "little men," which lie centrally along both sides of the brain. The posterior set of homunculi receives sensory input (touch, pain, vibration, position sense, temperature) from the body, and the anterior set is involved in motor, or muscular, output.[5]

When neurosurgeons repeatedly stimulate a specific brain area in the same patient, they always evoke the same exact memory, such as a song by Led Zeppelin. Neurosurgeons have also used recording electrodes to look at conditions under which brain cells will show electrical activity, or "fire." Single neurons in the hippocampus will fire when a patient is shown a specific face, such as Bill Clinton's or

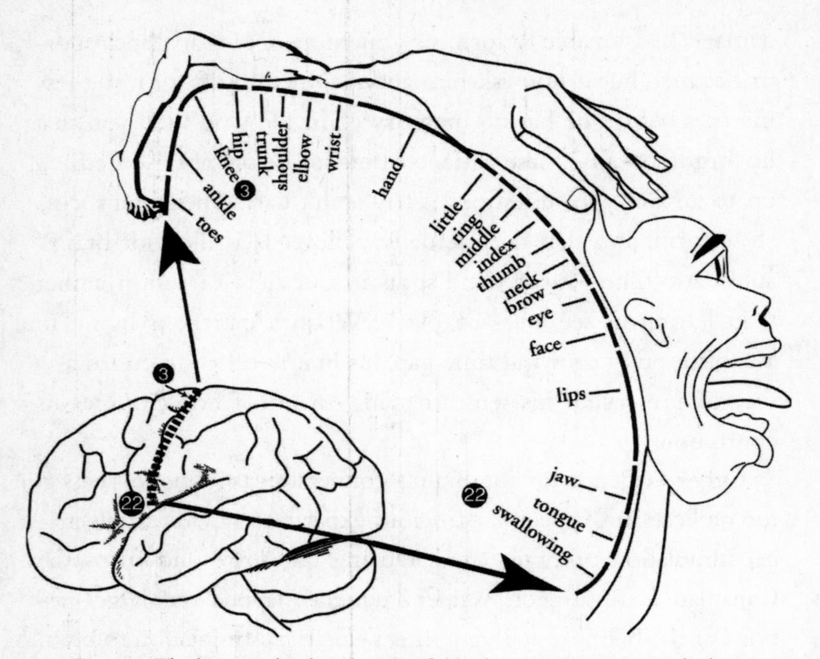

Figure 1. The homunculus, based on Penfield's diagram. Locations on the brain correspond to parts of the body such as the knee (3), and also to motor acts such as swallowing (22).

Jennifer Aniston's. The neuron appears to be committed to a specific face and will not fire for any other visual stimuli. Even more astonishing, it will fire when the subject is shown pictures from various angles or time periods, as long as they are of the same person. It is as though the brain cell is part of a network that has a concept of "Bill Clinton" or "Jennifer Aniston."

Reinforcement of the materialist view that the brain creates consciousness comes from brain imaging studies that reveal the brain areas activated during certain tasks.[6] Brain sections activated during addition are different from those used when conjuring up several words that begin with *r*. Brains also display different patterns when recalling something than when making it up, or lying. In fact, functional MRIs (fMRIs) have been proposed for lie detection. One can experience imaginary things as though they are real, and studies show that this activates the same brain areas as when we actually experience it. Brain imaging also has a characteristic pattern for being in a normal mood, depressed, manic, on illegal drugs, or sad.

THE HARD PROBLEM OF CONSCIOUSNESS

The brain plays a primary role in shaping our conscious experiences, but that is different from saying that it creates consciousness, although the materialist paradigm assumes it does. The philosophy professor David Chalmers called the inability to understand how something as nonmaterial as consciousness could arise from the brain "the hard problem of consciousness." The lack of any answer to the hard problem creates a missing link in the materialism paradigm.

The first approach to finding this link has been to look for something about the brain that is unique, different from other organs, so as to enable it to be the sole creator of consciousness. But brain cells have more in common with other cells than not. All mammalian cells have nuclei that contain the same genetic material as brain cells. Mammalian cells, except for sperm, have mitochondria that give them energy; they all have membranes that keep their interior contents intact while interfacing with other cells; and most have floating around in their membranes receptors for chemicals that influence cellular activity. In fact, many cells outside of the brain have receptors for neurotransmitters such as serotonin, the chemical messenger made famous by Prozac.

So what makes the brain different from other parts of the body? Brain cells differ from others by having electrical activity, called an action potential, that travels along their axons, which are the thin extensions that radiate out from the brain cell's body to form their connections with other cells. The only cells outside of the nervous system that operate by electrical activity are heart cells, which coordinate their activity with each other to create the muscular contraction of the heart. Unlike heart cells, brain cells are insulated from each other by fat to conduct electricity faster and to be more selective about which electrical activity affects them. If brain cells fire in a wavelike fashion, as heart cells do, this abnormal activity causes a seizure.

Another difference between brain and heart cells is that heart cells generate their electric current without stimulation, whereas brain cells require stimulation. A single heart cell isolated in a laboratory dish will have its own pulse, whereas a brain cell has electrical activity only in response to outside stimulation, such as an electric probe or another brain cell.

The majority of brain cell stimulation comes from other brain cells, but some comes from their interface with sensory systems. Unlike cells in most body organs, which connect only with adjacent cells, brain cells can have a variety of connections. For example, neurons in the front of the brain can connect with their neighbors, neurons in the back of the brain, and several neurons in between.

Because of differences in connectivity among brain cells, each cell has its own identity, analogous to people for whom a large part of their sense of self is derived from their relationships with others. There may be a huge overlap between the friends that I share with my best friend, but there are also differences between us in whom we know and how much we communicate with them.

Unlike cells in other organs, the connections between brain cells can undergo significant change. New connections are formed when we learn new things. Damage can also alter connections. If someone goes blind, some of the cells in the visual cortex can become recruited by another sense, such as touch, and they can strengthen that other sense beyond its usual capacity.

Whatever the source of human consciousness, it has to be complex. The brain certainly fits that bill. Most of its complexity comes from the enormous number of potential connections among brain cells. The brain has approximately 100 billion neurons, each of which has between 1,000 and 10,000 connections with other neurons at junctures called synapses. The calculated number of possible permutations, or combinations of brain activity, along this network exceeds the number of elementary particles in the universe.[7]

Another source of complexity is the tremendous variability in brain cells' connections with each other. Some brain cells connect through a direct electrical synapse, but the major means is through neurotransmitters, the chemical messengers that cross over from one brain cell to another at chemical synapses. There are more than a hundred neurotransmitters and multiple receptor subtypes for each neurotransmitter. Serotonin, norepinephrine, and dopamine are the most commonly discussed neurotransmitters because they are influenced by antidepressants, but they are not the most abundant.[8] Among the most ubiquitous neurotransmitters are GABA and glutamate. GABA decreases the likelihood that a connecting cell will fire, whereas glutamate is excitatory and increases the next cell's potential for firing.

The brain is a network of numerous on and off switches. In response to the cumulative input from other neurons, either neurons fire or they don't. A threshold has to be reached for a neuron to fire, but when it fires the output doesn't vary. It doesn't matter if the threshold was exceeded to a greater or lesser extent.

That all-or-nothing process reminded computer scientists of their binary computer code, which is written in 0's (off) and 1's (on). This led cognitive psychology to develop and embrace a "computational theory of mind" that regards the brain as functioning like a biological digital computer.

The computational theory of mind works well for understanding some functional aspects of the brain, but it does not do a good job of accounting for the experiential aspect of consciousness. In other words, it doesn't address how we can experience the tastes, smells, sights, or sounds of the world around us. How do we experience the color fuchsia, the smell of honeysuckle, or the taste of garlic bread? Where does the "I" that has those sensations come from or reside? How do our brains take separate sensory inputs of a multisensory experience and combine them so that we experience them as an external world, rather than as something occurring inside our heads?

One major difference is that computers can make decisions according to logic, but they can't have the experience of free will or the feeling of control over decisions about what to do next. They act out of a preset algorithm. Even if randomness is built into the algorithm to give it some unpredictability, it is not the same as free will. The concept of the brain as a biological network that operates like a computer suggests that we are just "zombies," or automatons, which is one of the model's major shortcomings.

If explaining how the material brain can create an apparently nonmaterial consciousness is the "hard problem," what is the easy problem? Scientists feel fairly confident that they'll locate the brain region responsible for conscious experience, as opposed to the brain regions whose activity remains unconscious. The temporal lobes were once proposed as the sites for conscious awareness because of their association with symbolism, which is important for much of our thinking. However, this can't be correct, because people whose temporal lobes were removed because of tumors or epilepsy still appear to be fully conscious beings. Another area of the brain was proposed by Francis Crick as a possibility primarily because its function is still unknown, but assigning function by default is unsatisfactory. So where conscious experience actually happens is still a very open question.

The easy question became more complicated when British and Belgian scientists studied the brain of a woman who had suffered severe brain damage in a car accident.[9] She was in a vegetative state, in which she could open her eyes but did not physically respond to sights, sounds, or being poked. Nonetheless, a functional MRI of her brain showed the same patterns as a normal person when she was asked to imagine playing tennis: the regions that were activated were those involved in movement. The areas associated with navigating space and recognizing places became more active when she was asked to imagine visiting rooms at her house. Prior to this study, people in a vegetative state were not thought to be conscious. Now scientists are revisiting their definition of consciousness.

THE NEXT STEP TO UNDERSTANDING CONSCIOUSNESS

Consider the following analogy: On one level we understand both music and consciousness by experiencing them. But trying to understand consciousness by investigating the gray matter in our skulls is like trying to comprehend music by dismantling CD players and analyzing their parts. This is why some neuroscientists are now collaborating with experienced meditators who have devoted their lives to deepening their experience of consciousness.

If what we want to understand is how CD players can play music, we need to know about physics, including that of lasers and computers. I believe that it is similarly impossible to understand the relationship between the brain and consciousness without looking to modern physics, which studies the essence of matter and energy and their interactions at the atomic and subatomic levels.

The brain is composed of atoms, and therefore the principles of quantum physics are operating in our brains, though most neuroscientists have yet to give significance to quantum principles. But a model that recognizes that quantum physics also operates in our brains might explain many of consciousness' unsolved mysteries. In other words, quantum physics just might provide the missing link that explains the relationship between something as nonmaterial as consciousness and something as material as the brain.

Most neurologists and psychiatrists are accustomed to interpreting fairly bizarre syndromes as a result of brain dysfunction, and many syndromes that are now accepted as genuine were at first misunderstood and often dismissed. I predict that psychic phenomena will eventually gain their acceptance as well if there are sufficient data. And if psychic phenomena are proven valid, these phenomena may serve as both clues to, and confirmation of, this developing model of brain function.

Chapter 2

DO YOU SEE WHAT I SEE? AN EXAMINATION OF THE EVIDENCE FOR TELEPATHY

Another dream-determinant that deserves mention is telepathy. The authenticity of this phenomenon can no longer be disputed today. It is, of course, very simple to deny its existence without examining the evidence, but that is an unscientific procedure which is unworthy of notice. I have found by experience that telepathy does in fact influence dreams, as has been asserted since ancient times. Certain people are particularly sensitive in this respect. . . . The phenomenon undoubtedly exists, but the theory of it does not seem to me to be so simple.

—C. G. JUNG IN *THE PRACTICAL USE OF DREAM ANALYSIS* (1934)

THE TERM *TELEPATHY* comes from the Greek *tele-*, meaning "distant," and *pathos*, meaning "feeling." Telepathy is the ability to feel from a distance and involves the communication from one mind to another without a visual, tactile, or auditory means. Sometimes it is referred to as "thought transference."

Telepathy has been reported to be more common in people who have a close social bond and high level of intimacy. This implies that the barriers people erect to intimacy may also be barriers to telepathy. One intimate relationship where the barriers are down is that between mothers and their newborns. In fact, mothers often report a sense of telepathic connection with their young children, especially infants who are too young to have formed their own sense of identity or personal boundaries.

The most common situations that evoke spontaneous telepathic communication appear to be when there is a threat to a member of the intimate relationship. The psychiatrist Ian Stevenson reported that 50 to 80 percent of such communications occur during a serious crisis.[1] Another common source is when there is a secret or something that the other is attempting to hide. There are many anecdotes of mothers knowing when their children are up to no good and of spouses who correctly develop suspicions when their partners cheat on them. But is this telepathy, or intuition, or the detection of subtle cues?

In a close-knit family, knowing the other's thoughts can happen even in mundane matters. Berthold Schwarz, a psychiatrist, kept a diary about coincidences that resembled telepathy within his family and wrote about them in a book, *Parent-Child Telepathy.*[2] One type of coincidence was when his children blurted out a comment as though in direct response to what the parent was silently thinking.

Many of us can relate to the above examples, but they don't get reported to telepathy researchers because they are not dramatic or clear-cut examples of telepathy. Of the reported spontaneous telepathic experiences, approximately 65 percent have occurred in dreams.[3] This suggested that the dreaming state may be conducive to receiving telepathic information, so telepathic dreams have played an important historical role in the investigation of telepathy.

The correlation between dreams and psychic phenomena has been noted for thousands of years. Ancient Eastern cultures believed that our consciousness traveled outside our bodies during dreams, thereby accessing information from other locations and realms. Many Jungian psychoanalysts believe in telepathy and regard dreams as symbol-laden clues to our unconscious mind, which is a portal for telepathic information.[4] Either way, telepathy's correlation with dreaming results from the shift away from normal waking consciousness. Dreams explore the usually inaccessible world of our unconscious, a world rich with contents of mysterious origin.

TELEPATHIC DREAMS

"Crisis telepathy" is one of the most commonly reported and well-documented categories of telepathic dreams, occurring when the sleeping person dreams of someone who is awake and in danger. In a "farewell dream" the loved one actually died when the dreamer simultaneously dreamed about the death. The explorer Henry Stanley had one of history's best-known farewell dreams. While in captivity during the American Civil War, he dreamed in detail of the unexpected death of his aunt in Wales. He later learned that she died at the same time of his dream.

Anecdotes such as Stanley's were an impetus to begin a serious scientific study of telepathy, so the Society for Psychical Research (SPR) was founded in 1882. SPR reported research findings from 149 cases of dream telepathy in the historic work

Phantasms of the Living. One typical story from the book is the following dream, which coincided with the time of the actual event and agreed with it in great detail:

> I saw father driving in a sledge, followed in another by my brother. They had to pass across a road on which another traveler was driving very fast, also in a sledge with one horse. Father seemed to drive on without observing the other fellow, who . . . made his horse rear. . . . I saw my father drive under the hoofs of the horse. Every moment I expected the horse to fall down and crush him. I called out, "Father! Father!" and awoke in great fright.[5]

Although many of the 149 case reports were compelling, the SPR research did not have an influential impact on the scientific community, primarily because case reports are anecdotal and don't happen under controlled conditions. They also aren't the type of phenomenon that one could reliably reproduce. Nonetheless, the following SPR conclusions are worth noting. Over half of the telepathic dreams concerned death, and another high number concerned an emergency. In most cases the dreamer and subject of the dream were related or friends. None of the dreamers had a history of psychic abilities before or after the dreams. And the dreamers were not prone to nightmares about others, so the dreams stood out as particularly disturbing. The combined findings led the researchers to question whether or not we all share this capacity for telepathic dreams, perhaps as an adaptive survival mechanism.

One question raised by the SPR research is whether dreams about the death of someone are common enough that statistically one would expect at least some dreams to coincide with an actual crisis or death. The SPR researchers attempted to address this question by sending out questionnaires to 5,360 people to ask them if they had had a vivid dream of the death of someone known to them within the last twelve years. Fewer than 4 percent of respondents reported that they had such a dream over the entire period of 4,380 days. Given the low frequency of such dreams, the researchers concluded that the odds of a farewell dream on the actual night of that person's death were against it being due to chance.

Another type of telepathic dream, the "shared dream," was heavily investigated by Hornell Hart.[6] Hart defined this as a dream "in which two or more dreamers dream of each other in a common space-time situation, and independently remember more or less of their surroundings, their conversation, and their interactions within the dream." How often shared dreams occur spontaneously is unanswerable because dreams are usually forgotten or are not discussed, particularly if they were not alarming.

People have actively tried to cultivate shared dreams. Sixteenth-century Sufi mystics, isolated in a monastery on the island of Rhodes, went to great lengths to have shared dreams as part of their spiritual practice. The Sufi master and his disciples recited a secret formula together and underwent group purification of their bodies, minds, and spirits before they slept in an enormous bed shared by everyone. Under these ritualistic conditions, they had the same dreams. But since what we do before sleep often becomes incorporated into our dreams, their shared dreams may have been due to their identical experiences prior to sleep rather than telepathy.

More convincing of telepathy are the spontaneous reports of shared dreams, such as the following: Dr. Adele Gleason dreamed of being deserted in a very dark wood. She was suddenly terrified that a man she knew would suddenly arrive and shake a tree next to her, an action which would cause the leaves to burst into flames. The dream occurred between 2 a.m. and 3 a.m. on January 26, 1892. Four days later she ran into the man who was in her dream. She wanted to tell him about her dream, but he interrupted and insisted on telling her about his dream, which he was certain was the same as hers. Indeed, at around 3 a.m. the same morning, he had dreamed that he had approached her in a dark wood, shaken a tree, and its leaves had fallen and burst into flames.

Because the context for the reported dreams was not included, one cannot say whether they were entirely independent from shared circumstances. However, the dreams are heavily symbolic and the parallels are very striking. Another interesting example comes from a couple that I met before I ever read about shared dreams. When Bob and Sally met, they recognized each other instantly, having been lovers in a series of shared dreams that they had had over a period of ten years before meeting. They married within three months after meeting in person.

LABORATORY RESEARCH ON TELEPATHIC DREAMS

After SPR's research, the next logical step was for research on dream telepathy to move to the laboratory for study under controlled conditions. The invention of the EEG, which measures brainwaves, and electro-oculogram (EOG), which measures eye movements, allowed researchers to know when someone was dreaming, because rapid eye movement (REM) sleep has distinct brainwaves and eye movements. Most laboratory work on dream telepathy was done at the Maimonides Dream Laboratory in

New York by Montague Ullman, M.D., and Stanley Krippner, M.D., during the 1970s and 1980s.[8]

Their largest telepathic dream experiment was done in 1971.[9] Ullman and Krippner asked two thousand attendees at a Grateful Dead concert to telepathically transmit a picture on display to Malcolm Besant, who was sleeping forty-five miles away in the Maimonides Dream Laboratory. The picture was of someone in the lotus position with brightly colored chakras, or energy centers, in alignment along his spine. Besant, a very successful English psychic and the grandson of a founder of the Theosophical Society, dreamed about a man "suspended in midair." He also saw "light from the sun . . . the spinal column."

Ullman and Krippner also did a series of experiments in which an "agent" in another room or building concentrated on a randomly chosen painting and attempted to transmit it to a dreaming subject. The subjects were awoken when the electrical recordings showed that they were dreaming. Their dream content was then compared to the painting. Often their dreams had a theme or content very similar to that of the painting. For example, if the painting was *The Last Supper*, it was counted as a positive result if the dreamer reported a dream about a feast or a dream about Jesus. The investigators declared an accuracy of 83.5 percent for those twelve experiments, but the criteria for a positive result were not very strict.

Over time the Maimonides procedure was revised. One protocol had the dreamers view between eight and twelve pictures after being awoken. They were asked to rank the pictures in order of relevance to their dream's content. It was counted as a hit if the "transmitted" picture was in the top half of the ranking and a miss if it was in the bottom half. This made statistical analysis of the data easier, but it also made the data less compelling than if they had been forced to choose one of the pictures. The changes in protocol over the years made it more difficult to do an overall analysis of their 450 trials than if they had kept the same format. However, a combined review, or meta-analysis, was done and it concluded that the overall success rate was 63 percent versus the chance rate of 50 percent; the odds were 75 million to 1 against the data being due to chance or guessing.[10]

Some of the most fascinating results in the Maimonides trials had to be counted as errors in the statistical analysis. For example, Alan Vaughn described a case where the dreamer picked up something from a staff member other than the sender.[11] Sol Feldstein was in charge of monitoring the equipment while the sender was concentrating on an art picture and the dreamer was asleep. The reported dream was about statues of women with their breasts exposed, which did not relate to the target picture. However, when Feldstein heard the results, he revealed that he had been reading an illustrated *Life* magazine article about topless bathing suits at the time.

In another example from the Maimonides research, the dreamer dreamed about the Northwest Mounted Police, which had nothing to do with the target picture. When the staff were discussing the results, the night monitor became embarrassed and confessed that he had fallen asleep on the job and had a dream about the Northwest Mounted Police.

Personal information about the researchers also sometimes contaminated dream content. Robert Van de Castle was a professor of psychiatry at the University of Virginia Medical School and was considered by Ullman and Krippner to be one of the best telepathic dreamers. He dreamed that Krippner's expense account statements showed him to be $25 short. When he and Krippner discussed the dream the next day, Krippner confirmed that he had not been reimbursed $25 of a business trip's expenses.

After the Maimonides research ended, forty-seven telepathic dream studies were conducted that differed primarily from their predecessors in that they allowed the dreamers to sleep at home rather than in the laboratory. This was considerably cheaper to conduct and allowed more trials, but it meant that the dreamers were not awoken immediately from their dreams or monitored by EEG/EOG. A meta-analysis was done of these 1,270 trials by British psychologists Simon Sherwood and Chris Roe from University College Northampton, England.[12] There was a hit rate of 59.1 percent, which was 9.1 percent over the chance rate of 50 percent. It is not surprising that the hit rate was lower than in the Maimonides trials because dreams are less likely to be remembered when subjects are not immediately awoken from them. However, because there were almost three times as many trials as in the laboratory and 9.1 percent was not that much lower than the 13 percent in the Maimonides trials, the odds ratio that the home studies were not a result of chance ended up higher than the laboratory odds ratio, at 22 billion to 1.

RESEARCH DURING ALTERED STATES OTHER THAN DREAMING

Altered states of consciousness have also been explored to see if they facilitate telepathic communication. One method of inducing an altered state is the ganzfeld procedure, which was named after a German word meaning "whole field," as in a whole field of consciousness. It involved a mild form of sensory deprivation that was originally developed in 1964 by psychologists Mario Bertini, Helen Lewis, and Herman Witkin to study altered states of consciousness.[13]

Charles Honorton, William Braud, and Adrian Parker applied this technique to the study of psychic abilities so that their subjects could focus their attention without competing external stimulation. They produced the ganzfeld state by affixing ping-pong ball halves over the receiver's eyes, which looked into a red light, and reducing auditory input by pink noise, which is white noise with the high-frequency components filtered out. A comfortable recliner and hypnotic-like suggestions relaxed the receiver. A sender looked at a picture and attempted to mentally transmit it over to the receiver while he or she was in the ganzfeld state. Afterward, the receiver was shown four pictures to choose the target from.

From 1974 to 2004, eighty-eight ganzfeld experiments were conducted, and 1,008 of the 3,145 trials were hits. Dean Radin, an experimental parapsychologist at the Institute of Noetic Sciences (IONS), reported in his book *Entangled Minds* that the combined hit rate was 32 percent, which is 7 percent higher than the 25 percent expected by chance.[14] Radin states that the odds against these results being due strictly to chance are astronomical: 29 quintillion (29,000,000,000,000,000,000) to 1.

The ganzfeld work was later refined by Adrian Parker and Joakim Westerlund. In these experiments one person views a film and telepathically sends the imagery to a person in the ganzfeld state. The precise time is kept track of so that the subject's reports from the ganzfeld state can be compared to what happens at that exact time in the film. The recordings of the reports are later superimposed upon the film in real time, as though the person is narrating the film. This allows for a second-by-second comparison of what was being "sent" and what was being "seen." One example of a hit was when the participant said, "Looks like something is being lifted, a tong, which holds something." The simultaneous five-second sequence in the film showed a handicapped man lifting an object using a type of tongs.[15] These results combine the compelling specificity found in the anecdotes with the controlled conditions of a laboratory, a combination that makes them an important contribution to the body of research.

Telepathic research was also done on people who could place themselves into a special meditative state. Some of the earliest research of this type came from an unlikely source: Upton Sinclair, the famous social activist and writer who exposed the horrific conditions of Chicago's meatpacking houses in his book *The Jungle* (1906), and who won the Pulitzer Prize for *Dragon's Teeth* (1942), a book about the rise of the Nazis in Germany. He became convinced of clairvoyance and telepathy through his wife, Mary Craig Sinclair, who was able to draw reproductions of sketches that were made by other people in remote locations. Sinclair's wife would purposely go into a deep meditative state or trance to obtain this information.

Sinclair conducted the research in a scientific manner, even though he was not a trained scientist. Despite concerns that he was risking his reputation, Sinclair wrote about these experiments in a book called *Mental Radio* (1930).[16] Albert Einstein wrote a preface to *Mental Radio* in which he stated that the book deserved an earnest consideration by the psychological profession and that he trusted the integrity of the author, whom he knew quite well.

Sinclair classified the 290 drawings by his wife into three categories: successes, partial successes, and failures. He considered 65 of the drawings, or 23 percent, to be successes, and 155 drawings, or 53 percent, to be partial successes. Although some of the "successes" are not very impressive, many are.

In one example of a success, Sinclair's brother-in-law had been instructed to sit in his home in Pasadena, California, select any object at random at 11:30 a.m. on July 13, 1928, draw a picture of it, and then concentrate entirely upon it for a period of fifteen to twenty minutes. At this precise time Mrs. Sinclair was lying on a couch in her home in Long Beach, forty miles away. She was in semidarkness with her eyes closed and engaged in a technique of mental concentration that she had been practicing for several years. When an image persisted in her mind, she considered it to be the correct one and wrote it down. On July 13 she wrote that she saw a table fork and nothing else, which was precisely what had been drawn.

RESEARCH ON TELEPATHY IN WAKING STATES

Telepathy has also been reported and studied in waking states. Sinclair's work on picture drawing attracted the attention of Whately Carington, a Cambridge University psychologist who conducted experiments in the 1940s that were more scientifically sophisticated than Sinclair's.[17] The experiments involved random selection of pictures, great precautions to prevent any possibility of fraud, and a judge who was unaware of the contents of the original drawings or their relationship to the ones drawn psychically. The judge had to match the original pictures to those psychically drawn solely by the similarity of their content. Out of 2,200 drawings, 1,209 were correctly matched to their counterparts. Statistically, these results would occur only 1 out of 30,000 times by chance alone. Because this research was done with 250 people who were not considered psychic,

Carington concluded that psychic abilities might be an attribute common to all of us.

The eminent psychologist William McDougall of Harvard University was also influenced by Mrs. Sinclair. During a visit, he asked her to describe the picture postcard (of an ivy-covered, Oxford College building) that was in his coat pocket. Mrs. Sinclair described a building with stone walls, narrow windows, and covered with green leaves. Convinced that psychic abilities deserved further study, McDougall started a program at Duke University.

Some of the most famous laboratory research on nondream telepathy was done at Duke University from 1929 to 1962 by J. B. Rhine with a deck of twenty-five ESP cards. The cards contained five different symbols (circle, star, square, wavy lines, and cross). In a typical experiment, one person randomly chose a card and tried to mentally send its symbol to another person. A series of experiments were published in 1940 in the book *Extra Sensory Perception After Sixty Years*. The most frequently cited experiments were conducted from 1933 to 1934 with Hubert E. Pearce Jr., who claimed he inherited his mother's psychic abilities. Out of seven hundred runs with the cards, his accuracy was 32 percent, which was 12 percent above the chance rate of 20 percent.[18]

Another early researcher on telepathy was René Warcollier, who became interested in telepathy after he had a series of telepathic dreams. Warcollier published more than fifty-six papers on telepathy and clairvoyance during his time at the International Metapsychic Institute in Paris.[19] In 1948 he published the book *Mind to Mind*, which contains many pictures of both the transmitted target and the picture drawn by the receiver.[20] Many pictures show targets that are extremely similar to their corresponding pictures. One picture is of eyeglasses. It looks almost exactly like the eyeglasses target, even though the drawer didn't recognize what the object was.

Warcollier attempted to understand the differences between pictures and their targets. He saw recurrent themes in the types of errors made. He also tried to discern the properties of objects that made them more successful targets. His conclusions can be summarized as follows:

1. Emotional states are more easily perceived than intellectual material.
2. Intellectual images are subject to certain types of distortion. A square might be drawn as two or more disjointed right angles. Concentric circles might be drawn as nests of detached arcs.
3. The drawing's accuracy does not require the receiver to recognize the target.
4. The relationships between an object's individual parts may be drawn correctly, but the parts may be poorly defined.
5. Similar to dream imagery, several images may be condensed into one image.
6. Targets that move appear to be more powerful.
7. Targets are better if there is a sharp contrast between figure and ground.
8. Targets are more powerful if they are composed of parts that are equally divided, or are repetitive.
9. The perception of a target's color can occur independent from its form.
10. Imagination can interfere with correctly perceiving the target.

Warcollier's conclusions might just be attempts to find a meaningful connection between the drawings and the original targets. But if telepathy is real, they may be valid observations that tell us something about the brain's portal for telepathic communication. Several of the conclusions suggest that the brain's circuitry for dreaming is involved. The notion that emotional stimuli work better than nonemotional ones, a finding the SPR also reported, corresponds with emotional processing's use of the brain circuitry for dreaming to resolve conflicts and emotions during dreams.[21] Another observation—that form and color can be separated in telepathic imagery—also describes what happens in dreams, which frequently are in black and white. Warcollier also found that the rational mind interfered with accuracy, and in dreaming the rational mind is absent or muted. Also, the perceived images were often condensed, as images are in dreams.

ANIMAL TELEPATHY

Although this book is about human consciousness and psychic abilities, any chapter on telepathy would seem incomplete without reference to animal telepathy. Rupert Sheldrake, a Cambridge University–trained biochemist who subsequently turned his research skills to animal behavior, has conducted surveys of randomly selected American and British animal owners to find out how many of them believe that their animals can telepathically respond to their thoughts. An average of 48 percent of dog owners and 33 percent of cat owners believed that they do. Also, many of the people who work professionally with dogs and horses believe in animal telepathy. Barbara Woodhouse, the famous British dog trainer, said, "You should always bear in mind that the dog picks up your thoughts by an acute telepathic sense and it is useless to be thinking one thing and saying another; you cannot fool a dog."

But is there any research evidence? Some of the earliest research on animal telepathy was done in the 1920s by Vladimir Bekhterev, a Russian neurophysiologist.[22] He became intrigued after witnessing a circus act in St. Petersburg in which a fox terrier seemed to respond to the mental commands of his trainer, Vladimir Durov. Durov described his method as looking into the dog's eyes while he visualized the task that he wanted the dog to do, such as fetch a specific item from a table. Careful that he was not giving out any subtle clues, Bekhterev carried out a series of successful trials with the fox terrier. He then conducted trials with his own dog, which were also successful. He found that it was possible to telepathically communicate even when he and the dog were separated by screens, or when the dog was blindfolded.

In his book *Dogs That Know When Their Owners Are Coming Home and Other Unexplained Powers of Animals*, Rupert Sheldrake provides numerous accounts of animals who appear to know telepathically when their owners have made the decision to return home.[23] His research was documented by simultaneously filming the owner and his or her pet in their separate locations. A characteristic study shows a dog or cat who goes to the front door at the same time the owner has picked up his or her briefcase and started walking toward the office door. The time of day when the owners headed home varied, but the behavior of the pet still occurred in sync with the owner. If the owner changed his or her mind and stayed at work, the animal immediately left the front door to return to a more usual spot. Sheldrake also has a database of more than 108 reports of people witnessing dogs that appeared to respond with distress at the exact time of the distant accident or death of their human companion, and 51 accounts of cats that did so.

EXPERIMENTS ON COUPLED CONSCIOUSNESS

If people experience concurrent and identical changes in brain activity when only one of the pair is given a stimulus, this would suggest a coupling of consciousness. Research by Marilyn Schlitz and Dean Radin was done on twenty-six pairs of volunteers, whose brainwaves were measured by EEG simultaneously while they were in separate rooms. If one of the pair was shown video images, the other pair member had corresponding EEG changes, as though the images were shown to him or her as well. Some of this work showed positive results even when the pair were strangers, provided they had spent an hour or so together to form a bond, which can happen with effort within this relatively short period of time.

Similar studies were done by Jacobo Grinberg-Zylberbaum and Julieta Ramos of the Universidad Nacional Autonóma de México.[24] Thirteen pairs and four groups of three volunteers in separate rooms and Faraday cages (metallic rooms that provide electromagnetic shielding) were asked to feel each other's presence.[25] The results showed that their EEGs began to synchronize with each other when they did this. Also, the electrical activity within and between the cerebral hemispheres of subjects began to synchronize and showed an increased coordination between the right and left hemispheres during their attempt to be in telepathic contact. Subjects in one session said that they had a sense of "having blended" with the other member of the pair, and their EEG patterns were nearly identical.

Functional MRI was used by several researchers from Bastyr

University and the University of Washington, Seattle, to look at coupled consciousness. One member of a pair of colleagues had a checkerboard pattern flashed in front of his or her eyes, which caused increased activity in the visual cortex. The other member of the pair simultaneously had his or her brain scanned. The male colleague did not show any changes when the female colleague was shown the checkerboard pattern, but the woman did when the man was shown it. This is an interesting finding since telepathy is reported more commonly by women.

Other research suggests that the brain may not be the only body part involved in telepathy. A more primitive "brain," or neural network, in the gut is thought to be involved in emotional responses, or "gut reactions," which can occur in the gut independent of the brain. The gut contains more than 100 million neurons and is the only organ other than the brain with such a complex neural network.[26] Electrodes on the skin of the abdomen can measure activity in the gut, just as the electrocardiogram (EKG) can measure one's heartbeat from the skin on the chest. The recording is called an electrogastrogram (EGG), and a typical rhythm in the gut is three "beats" per minute, but it changes in response to strong emotions.[27]

Dean Radin and Marilyn Schlitz used the EGG to see if individuals could be in telepathic communication at a "gut level."[28] Twenty-six pairs of adult friends or relatives participated. Each pair had a sender and a receiver. The EGG of the receiver changed more when the sender was experiencing strong emotions than when the sender was emotionally neutral. This was statistically significant for both positive and negative emotions. Although one cannot conclude that the changes in the receivers' guts were entirely independent of their brains' unconscious activity, this study demonstrates another measurement of linked responses between senders and receivers.

WHAT THE DATA ON TELEPATHY MAY MEAN

The research studies discussed here give results that are statistically significant by scientific standards. The likelihood is that at least one of these positive results represents a true case, which by James's criteria is sufficient to say that the phenomenon exists. There actually are many more positive studies than mentioned here; this brief review offers only a representative sampling of the research done. Not all published study results have been positive, but a significant number have been. This is not due to what is known as a "file drawer" effect, which refers to the skewed presentation of data that happens when people publish only positive results.

In *Entangled Minds*, Dean Radin shows that the file drawer effect can easily be detected by plotting the results of all the published studies in a graph. Enough graphs of published studies have been done for other areas of research that there is an expected shape for the plotted graph. If someone has thrown out undesirable data, it shows up as a characteristic distortion in the graph's shape. The data look hand-picked, rather than au naturel. Radin's analysis showed that a file drawer effect was not the case for telepathic research.

From a qualitative standpoint, the laboratory studies are less impressive than the anecdotes, but the experiments also lacked the features mentioned in the SPR conclusions as conducive to telepathy. The laboratory research did not involve an emergency or crisis with a loved one, which is something that would be unethical to experimentally produce. Instead they had people focus on emotionally neutral subjects such as art or ESP cards. Also, telepathy has been reported to be more common in people who are creative. Perhaps telepathy, like creativity, usually happens spontaneously and is difficult to turn on with the flick of a switch.

Another criticism of this research is that scientists who believe in psychic phenomena tend to have positive results, whereas skeptics tend to have negative results. The believers respond to this with an explanation called "experimenter psi." In other words, if one's intention can influence outcome, which is a form of psychokinesis, then experimenters' beliefs will influence their own experiments. If that is the case, skeptics may never be able to reproduce the results of the believers. The problem with this argument is that skeptics in other areas of research have been able to prove themselves wrong. Another explanation for this difference in results may be that most skeptics don't have any interest in doing research on psychic abilities, so they have far less data for it to be statistically significant. A difference of 10 percent is more meaningful if there have been one thousand trials than if there have been ten. Think of a coin toss. Getting 6 heads and 4 tails is more likely than 600 heads and 400 tails. Another factor is that skeptics also may create a less conducive environment.

Although we have statistical evidence suggestive of telepathy, this leaves us with two basic ways of interpreting these data. These positive results may be examples of synchronicities, coincidences that appear to have meaning because they greatly defy the odds of happening. Alternatively, the results could be evidence that individuals are able to have their consciousnesses in sync with or coupled with others'.

There is no doubt that coincidences occur and that some of the data may represent just that. But the data presented in this chapter do at least suggest the possibility of coupled consciousness. The studies using EEGs and fMRIs are very consistent with this hypothesis. And if telepathy is possible, it would be evolutionarily advantageous, especially during a crisis. Perhaps telepathy evolved long ago in animals, became buried during the development of our analytical brains, and now surfaces primarily in our dreams.

Chapter 3

TWO HEARTS BEAT AS ONE: IDENTICAL TWINS AND COUPLED CONSCIOUSNESS

Our separation from each other is an optical illusion of consciousness.

—Albert Einstein

The debate about telepathy versus synchronicities is particularly important when looking at the data on identical twins, about whom anecdotes of telepathy, coupled consciousness, and synchronicities are both abundant and remarkable. The eminent British scientist Francis Galton published a short article on twins in 1876 in which he commented that there were twins who, while in the company of each other, were witnessed to "make the same remarks on the same occasion" or "begin singing the same song at the same moment."[1]

But even when identical twins are apart they often act in the same way. There are countless stories of identical twins who unknowingly bought each other the same gifts and presented them to each other at the same time. The SPR researchers noted that there was a higher incidence of telepathy between twins than between other sibling pairs. And, according to Guy Playfair, author of *Twin Telepathy*, as many as 30 percent of identical twins appear to experience telepathic interconnection.[2]

RESEARCH ON TWIN TELEPATHY

Some of the earliest research on twin telepathy was done by a twin, Professor Horatio Newman, head of the zoology department at the University of Chicago. He had what he considered telepathic experiences with his twin brother and published a book called *Twins and Super-Twins* in 1942 that included a section on telepathy.[3] He discussed identical twins who were both conservative biologists. They were mystified by the way in which they could communicate with each other without any verbal exchange.

Robert Sommer, Humphry Osmond, and Lucille Pancyr interviewed a total of fourteen pairs of twins and seven single members of a twin pair to see how many of them reported experiences of telepathy.[4] Twelve out of the thirty-five participants believed that they could communicate telepathically with their twin. They made statements such as "We both think the same things at the same time," "I can tell what her feelings are," and "When my twin goes out, I can imagine what he is doing and see the place, like right now, even if I've never been there or seen the place described." As in the SPR conclusions, telepathy happened frequently between closely connected twins during crisis.

The term "crisis telepathy" was coined after several dramatic accounts such as the following: Martha Burke felt as if she "had been cut in two" one day in 1977 when a searing pain crossed her chest and abdomen. Hours later she discovered that her twin sister had died in a plane crash halfway across the world. Similarly, in July 1975, Nita Hurst's left leg became agonizingly painful as bruises spread spontaneously up the left side of her body. She later discovered that her twin, Nettie Porter, had been in a car crash at the very same time four hundred miles away.

Unfortunately, there are not a lot of laboratory data on identical twins and telepathy. The British parapsychologist and skeptic Susan Blackmore tested twins in separate rooms for telepathy.[5] First she asked them to draw whatever came into their minds. They often drew the same things. However, when she asked one to draw an object and telepathically transmit it to the other twin, the results were disappointing. Blackmore concluded that these twins weren't clairvoyant or telepathic, but that they thought alike.

Frank Barron, at the University of California at Berkeley, conducted a study in 1968 on twenty-six pairs of identical twins.[6] The twins were in separate rooms and one of the twins was shown films with strong emotional stimuli while the other twin's skin conductance, heartbeat, and respiration were measured. Only one pair of twins showed any positive results. However, Barron realized in retrospect that his testing conditions were far from ideal. Many of the subjects were apprehensive to begin with, they did not get a chance to know the experimenter, and the laboratory setting was very unfavorable for relaxation. So if the measured physiological parameters reflected the twins' heightened state of arousal, like the "white coat effect" that raises people's blood pressure at the doctor's office, it would be difficult to detect changes in the direction required to get a positive result.

A study by Edward Charlesworth at the University of Houston had one member of a twin pair look at a picture.[7] The other twin was placed in a comfortable setting, asked to daydream, and then asked to rate six pictures on a scale from most to least likely to be the target picture. Positive results for the identical twins occurred in seven out of twenty pairs, less than the chance rate of ten out of twenty. However, for nonidentical twins the results were fifteen out of twenty. Puzzled by the data, the researchers gave them personality tests and found that the highest telepathic results occurred if the twins were extroverts rather than introverts.

Being genetically identical may aid telepathy between twins, but it doesn't appear to be the most important variable. Since extroverts, by nature, thrive in the company of others, they tend to form more social bonds. Other data have shown that social bonds facilitate telepathy. Perhaps the extroverted twins were more bonded and intimate with each other. Another possibility is that the brain chemistry of extroverts facilitates telepathy. It also may be easier for extroverts to relax in laboratory settings.

Subsequent research in the 1980s by a French doctor named Fabrice-Henri Robichon measured subjects for extroversion so as to only test extroverts for telepathy.[8] His set of Zener ESP cards had five symbols: blue wavy lines, green stars, black squares, yellow circles, and red crosses. Though he tested only one set of twins, his results were markedly better than chance, which was 20 percent. He tested them five times and their scores were 64, 92, 72, 80, and 88 percent. The sample size was very small, but the high difference above chance suggests that these twins were more capable of telepathy than others.

Another approach has been to simultaneously measure EEGs of identical twins in separate rooms to see if they have coupled consciousness. The first study was done in 1965 by two Philadelphia ophthalmologists, T. D. Duane and Thomas Behrendt. The EEGs showed striking correspondences in two out of the fifteen pairs. When one twin closed his or her eyes, both twins demonstrated the usual increase in alpha waves that occurs with closing the eyes, even though the other twin's eyes were open. These results appeared in *Science* in 1965.[9]

The reaction to the ophthalmologists' article was extreme. *Science* is a very elite journal read primarily by professional scientists. The journal's editors received many complaints about this article. The major ones were that the sample was too small and the details of the study's design were too vague. More than thirty years later it was revealed that the study was funded by the CIA and still classified at the time of publication. That may explain why it was so vague and why the researchers didn't defend the results by providing more information.

IDENTICAL TWINS RAISED APART

The University of Minnesota research on identical twins raised apart yielded some startling findings, many of which are discussed in Dr. Nancy Segal's book *Entwined Lives*.[10] Sixty-eight cases have been extensively studied. When reunited, these twins often felt as though they had known each other their entire lives. They had an immediate ease of communication between them, as though they had been in contact all along. Their similarities in personality and appearance were remarkable but not unexpected because of their identical genetics. What startled the researchers were the many shared life details between the twins, which defied the odds of chance and conventional understanding. The following examples illustrate that point.

The "Jim twins" had been separated at four weeks and were apart for thirty-nine years. Both were named Jim, married a woman named Linda, divorced, and then married another woman named Betty. However, one Jim was on his third marriage. They both had childhood dogs named Toy and sons named James. One son was James Allen and the other James Alan. They both had been firemen and sheriffs. Both bit their nails, suffered from migraines, smoked Salem cigarettes, and drank Miller Lite beer. Each was six feet tall and weighed exactly 180 pounds, but they wore their hair differently. Among the most remarkable shared details was that both had had a compulsion to build a circular white bench around a tree in their yards during the time right before they met. Also, they both had regularly driven light blue Chevrolets to Pass-a-Grille Beach, Florida, for family vacations. They also both left love notes to their wives throughout their houses. Their facial expressions, IQs, habits, brain waves, and handwriting were nearly identical. To top it all off, they died from the same illness on the same day.[11]

Bridget Harrison of Leicester, England, and Dorothy Lowe of Burnley, Lancashire, England, were reunited in 1979 after thirty-four years apart. When they met, both wore seven rings, two bracelets on one wrist, and a watch and bracelet on the other. One's son was named Richard Andrew, and the other's was Andrew Richard. Both had a cat named Tiger, had stopped taking piano lessons at the same age, and had kept a diary in 1960. They had chosen exactly the same brand and color of diary and left the same days blank during the year.

Barbara Herbert found her lost twin, Daphne Goodship, after forty years of separation. Both grew up outside of London, left school at age fourteen, fell down stairs and injured their ankles at age fifteen, worked in local government, met their future husbands at the town hall dance at age sixteen, miscarried in the same month, and then gave birth to two boys and one girl. Both tinted their hair auburn when they were younger, were squeamish about heights and blood, preferred cold coffee, and burst into laughter readily. They had a habit they independently called "squidging," which was to push their nose up with the palm of their hand. They also were dressed alike when they met: both wore cream-colored dresses and brown velvet jackets.

Another set of reunited twins discovered while unpacking that they used the same shaving lotion (Canoe), hair tonic (Vitalis), and toothpaste (Vademecum). They both smoked Lucky Strikes and later mailed each other identical birthday presents.

Oskar Stohr of Germany and Jack Yufe of California were separated after their birth in Trinidad. Yufe was brought up as a Jew in Trinidad, while Stohr was raised in occupied Czechoslovakia and went to a Nazi-run school. Both wore a short, clipped moustache, stored rubber bands around their wrists, read magazines back to front, and had the habit of sneezing loudly in public to attract attention.

Adriana Scott and Tamara Rabi were born in Mexico and grew up within twenty-five miles of each other outside New York City but didn't reunite for twenty years. Adriana went out with a boy named Justin Lattore, whose friend subsequently set him up with a girl named Tamara. Justin was shocked at Tamara's resemblance to Adriana. Then he discovered that Tamara, like Adriana, had been adopted from Mexico and had her same birth date. He told them about each other. When they reunited they both wore bubble jackets, but one was black and the other was purple.

Clearly these coincidences defy the odds. To calculate the exact odds would require a great deal of research into the frequency of specific names during their years of birth, the frequency of divorce and remarriage for their locations and generations, the number of certain clothing items that had been sold, and the incidence of various odd details that wouldn't have statistics available. However, to illustrate how one would calculate the odds if they had these numbers, I derived some conservative hypothetical numbers. Let's say the odds of a boy being named Jim are 1 in 50, the odds of marrying someone named Linda are 1 in 100, the odds of divorce are 1 in 2, and the odds of remarriage to someone named Betty are 1 in 300. The odds of that combination are derived by multiplying the odds of each component, and equal 1 in 3 million. Even that is too conservative for the Jim twins, because it doesn't include the circular fence around the tree, or being firemen, or vacationing at the same beach, or dying on the same day, or having sons and dogs with the same names.

WHAT KNOWN FACTORS INFLUENCE THE LIKENESS OF IDENTICAL TWINS?

I've personally known fifteen sets of identical twins and three sets of identical triplets during my life. They are one reason I have been so curious about what makes us who we are. How much is due to our genetics, and how much is due to our environment? And how do these two influences interact? Each of the twins and triplets I've known was raised with his or her identical sibling(s) in the United States, where the need to be recognized as an individual is very strong. Their need to establish their own identities made them strive to be different from each other. When identical twins grow up in entirely different households (sometimes different cultures), none of their choices stem from reactions to being a twin. So twins raised apart have the opportunity for the same level of self-expression as singletons.

Researchers have used the twins-separated-at-birth data to justify the idea that genetics plays a far greater role in who we are than we realize. However, the problem is that the human genome isn't complex enough. The Human Genome Project predicted that humans would have more than 100,000 genes. It was a surprise when the entire human genome turned out to contain only around 35,000 genes. Given that the mustard plant has 25,000 genes, a life-form's complexity is not proportional to its number of genes. And the overlap between human and banana plant genomes is 43 percent, while the overlap between chimpanzees and humans is 98.5 percent. Obviously, the number of genetic differences between two distantly related humans is far fewer. Estimates say that only 3 million pairs of nucleotides (the basic components of DNA) distinguish each of us from any other person on the planet. This may sound like a lot, but the human genome contains roughly 3 billion nucleotide pairs. So the distinguishing DNA is only one thousandth of the total. The complex similarities between separated twins simply cannot be accounted for by science's genetic model.

Ninety-eight percent of our chromosomes contain what has been called "junk DNA." It was considered "junk" because, although it is made up of the same DNA bases as genes, it cannot be translated directly into the manufacture of proteins, which is how genes produce their effects. Mutations within "junk DNA" are common and don't seem to have consequences or to get eliminated by the evolutionary process of natural selection. So "junk DNA" was considered just a placeholder, or framework, for the important DNA. But now that the human genome has been found to be too small to explain human complexity, researchers are turning to "junk DNA" for answers. These answers would require a mechanism different from protein production, however.

A field called epigenetics is the science of what regulates genes, or causes them to turn on and off. Genes are turned off by methylation, which is the addition of a small carbon-based molecule to the backbone of DNA by a substance called a methyl donor. Environmental factors can cause this methylation of genes at any time in life, including during critical stages of development in the womb.

Randy Jirtle and Robert Waterland of Duke University fed pregnant mice a diet rich in methyl donors, which are found in onions, garlic, and other foods.[12] The pregnant mice had a gene known as agouti, which made them yellow, fat, and prone to diabetes and cancer. The methyl donors in their diet attached to the chromosomes of their developing embryos and modified the expression of the agouti gene in them. As a result, the mouse offspring were slender, brown, and lived to a healthy old age. Jirtle said, "Before, genes predetermined outcomes. Now, everything we do, everything we eat or smoke, can affect our gene expression and that of future generations." Identical twins, even if they were separated at birth, usually had the same epigenetic influences in the womb. But after birth they are exposed to many different epigenetic factors, especially if they were separated. Like genetics, epigenetics can't explain their uncanny similarities.

The explanation also can't come from the shared hard wiring in their brains. The brain has a "plastic" quality, meaning it is highly influenced by its environment. Its connections are continually changing in response to what is learned, reinforced, or ignored. There is so much variation in the wiring among individuals that brain surgeons test areas in their patients' brains before cutting in order to avoid unnecessary interference with the most critical sections. Such a highly changeable and environmentally influenced system makes many of the commonalities among the separated twins even more remarkable.

Richard Rose, professor of psychology and medical genetics at Indiana University in Bloomington, has studied personality in more than seven thousand sets of twins and believes that environment, shared and unshared, plays a larger role in personality than genetics. He also examined a factor besides genetics and upbringing that explains why some twins are more similar than others: the timing of separation of the developing embryo into identical twins. Identical twins result when a single egg is fertilized and separates into two developing embryos shortly afterward. This timing doesn't apply to nonidentical twins, or fraternal twins, because they result from fertilization of two separate eggs. Genetically, they are no more alike than other siblings, but they share more influences in the womb with each other than they do with their singleton siblings.

Twins formed from a single egg will always have identical genetics, but the degree to which they share the same womb environment is determined by when they separated into two embryos. If the separation occurs in the first four days of pregnancy, each twin has his or her own placenta, chorionic sac, and amniotic sac, like nonidentical twins. If the split happens between the fifth and eighth days, identical twins have separate amniotic sacs but shared placentas and chorionic sacs. Twins whose split was between eight and twelve days share all three. They are so close that their umbilical cords can get entangled. When the developing embryo splits after twelve days, they are conjoined twins, which means that their bodies did not fully separate.

Rose studied identical twins whose time of separation could be estimated by the visual examination of their afterbirth for differences in the sharing of chorionic sacs, placentas, and amniotic sacs. He found that the earlier the egg separates, the less alike the twins are in personality. Some have attributed the greater similarities in later-split embryos to a more closely shared womb environment, but another possibility is that it may be a result of the higher number of days they were still "one individual."[13]

WHY THE TELEPATHY FINDINGS DEFY CONVENTIONAL EXPLANATIONS

The evidence for coupled consciousness or telepathy between identical twins is dramatic but primarily anecdotal. The shortage of compelling evidence from laboratory studies on twins may be a result of the paucity of studies as well as how they were conducted. None of the laboratory settings involved crisis or danger to the other twin, which would have increased the chances of crisis telepathy. Some settings were not relaxing, which decreased the likelihood of noncrisis telepathy. Other factors associated with positive outcomes were missing. These include testing only extroverts, ensuring belief in telepathy (on the part of both the experimenter and the subjects), and being in a dream, meditative, or ganzfeld state. So the twin laboratory studies have not been optimal for demonstrating telepathy.

However, the data from the separated twins reveal parallels between their lives that are decidedly far greater than chance. The parallels obviously don't come from a shared environment. The twins have identical genetics and, to differing degrees, shared environmental factors in the womb. One can explain some of the similarities by genes (perhaps their similar tastes in cigarettes or colognes, similar interests or careers, looks, IQs, etc.), but there are other parallels (such as building a circular white fence around a tree) that do not have a genetic explanation. If such things were genetic, it would suggest that we are just biological machines. But if that were the case, why aren't identical twins even more alike? And how could genes possibly be coding for this, given the size of the human genome?

So what are the possible explanations for the uncanny parallels? Some things, like marrying people with the same name, seem to fit the category of synchronicities better than telepathy, whereas the simultaneous experience of pain when the other twin is being harmed seems to fit the category of telepathy. There are many things in the gray zone, such as buying each other the same gift. The twins who were both named Jim by their adoptive parents may have something about them that makes them look like a "Jim." Their names could also just be a coincidence, since Jim is not an unusual name. But the name, in combination with so many other similarities, becomes part of a mega-synchronicity.

Rather than getting caught up in determining which is which, another way of looking at synchronicities and telepathy is that they may be manifestations of the same underlying scientific principles of how the universe works. As a scientist, I don't regard these phenomena as supernatural. When something is labeled "supernatural," that only means that we haven't figured out the operating natural laws. Telepathy and synchronicities are important clues to a greater understanding of the mysteries of the universe and human mind.

Chapter 4

CLAIRVOYANCE: THE ABILITY TO SEE REMOTELY

Every man takes the limits of his own field of vision for the limits of the world.

—Arthur Schopenhauer

WHEREAS TELEPATHY IMPLIES coupling one's consciousness with that of another conscious being, clairvoyance is visualizing distant or hidden images as though one had a psychic telescope, periscope, or camera. Clairvoyance (sometimes called remote viewing) has reportedly found oil, mineral deposits, hidden treasure, and missing people. It has also been reported to diagnose medical and mechanical problems by seeing inside ill people and malfunctioning machines.

Telepathic-like experiences (such as thinking something at the same time as someone else) are more common than psychically seeing what is behind an unopened door, inside an envelope, or halfway across the world. However, reports of clairvoyance are not rare. Surveys across several nations show that one-third to one-half of the general population thinks they've experienced telepathy, and one-fifth reports experiences akin to clairvoyance.[1]

There are at least two different types of clairvoyant experiences. For some the images arise from their unconscious just as telepathic thoughts do. The image can come spontaneously in a flash, or it can evolve over time while the person concentrates on retrieving it. In the other type, clairvoyant experiences occur in what are called "out-of-body experiences" (OBEs). During an OBE people no longer feel that their consciousness is confined by their body; the visual perspective they experience would be impossible if it were. They may see something remote or from a different angle than their head would allow (such as seeing the room from the ceiling).

CLAIRVOYANCE AND MEDICAL DIAGNOSIS

One of the most famous clairvoyants of all time was the "sleeping prophet," Edgar Cayce, who was born in 1877 and died in 1943. Cayce's family were prominent Kentucky tobacco farmers, and his grandfather was locally known as a psychic. Like many psychics whose abilities appeared after brain injury, Cayce had such an accident. When he was three years old, he fell off a fence post onto a board with a protruding nail that not only punctured his skull but also entered his brain. Reportedly, he escaped any serious consequences.

Starting in 1901, Edgar Cayce did more than 14,000 clairvoyant readings based solely upon the client's name and address. He told his secretary about his visions while still in a self-induced OBE. Of these readings, more than 9,400 concerned medical diagnoses and treatment recommendations. He originally wanted to live a normal life as a professional photographer, so his readings were just a hobby for the first ten years. But his hobby helped so many people that it became his "calling" and continued the rest of his life. Although he never gained financially from his readings, others profited from his predictions about the stock market and suggestions about where to drill for oil.

Cayce's fame became widespread in 1910 after meeting Dr. Wesley Ketchum, a reputable homeopath who decided to get a reading for himself. Dr. Ketchum had a condition he had self-diagnosed as appendicitis, which was confirmed by other doctors. He was scheduled for surgery, but Cayce gave him the diagnosis of a spinal condition that was impinging upon a nerve. After Cayce's diagnosis was confirmed by successful spinal manipulation, Dr. Ketchum consulted with Cayce on his most difficult cases and submitted a paper to the American Society of Clinical Research that declared Cayce a "medical wonder." The *New York Times* spread the news. On October 9, 1910, its headline read: "Illiterate Man Becomes a Doctor When Hypnotized."

Among Cayce's visitors were Woodrow Wilson, Henry Wallace, Nikola Tesla, Thomas Edison, and many industrialists, bankers, and doctors. The psychologist Gina Cerminara spent a year studying records of his readings and wrote her conclusions in *Many Mansions.*[2] She found that many readings were extremely accurate and the clients improved after following Cayce's recommendations. Another of the scores of books about him is *Edgar Cayce in Context: The Readings: Truth and Fiction* by Kenneth Paul Johnson.[3] Considered to be both objective and open-minded, Johnson concluded that Cayce had a genuine gift. Cayce was also labeled an authentic mystic by Baba Faqir Chand, an Indian guru known for his outspoken criticism of many gurus, prophets, and mystics whom he regarded as duping millions of people. Cayce's sons, Edgar Evans Cayce and Hugh Lynn Cayce, compiled their own study of their father's work. They found only 200 flawed readings out of 14,246.[4]

Edgar Cayce is considered the "father of holistic medicine" because his recommendations form the basis for much of the holistic medicine still practiced. Most traditional Western doctors consider his treatments, such as colonic therapy (enemas) for detoxifying the body, to be quackery. But millions of people follow and swear by practices introduced by Cayce, and some have gained scientific validity. For example, he advocated eating almonds, which have been proven to have many health benefits. He also saw the importance of spiritual and psychological issues in people's health.

Like Cayce, medical intuitives psychically diagnose illness in patients without information from tests or interviews. Medical intuitives often claim to have no other psychic abilities, whereas Cayce later branched out into more esoteric and controversial areas, such as readings about past lives and ancient cultures. And Cayce made medical diagnoses while in a deep trance, but most medical intuitives do not operate this way.

The physician Norman Shealy researched medical intuition in the 1980s with the medical intuitive Caroline Myss. Their study involved fifty patients who individually sat in Shealy's consultation room while Myss made diagnoses based upon their names and birth dates from her office twelve hundred miles away. She felt that the physical distance was an advantage because personal connections with patients sometimes blocked her ability. The results are described in their book *The Creation of Health*.[5] Shealy reported that Myss had an overall accuracy of 93 percent. Examples of diagnoses Myss and Shealy made in common were schizophrenia, migraine headaches, myofascial pain, depression, sexual problems, venereal herpes, back pain, anxiety, wasting of the brain or Alzheimer's, and epilepsy.

RESEARCH ON REMOTE VIEWING

Remote viewing differs from remote medical diagnosis because the target is an object or a scene. In remote medical diagnosis the psychic may be tapping into someone's conscious or unconscious knowledge of what is wrong. In order to establish remote viewing as distinct from telepathy, no one else should know the target. But many remote-viewing experiments were intentionally set up with someone concentrating on the target in order to give the remote viewer the maximum opportunity for correct identification.

A famous study on OBEs and clairvoyance was done by Charles Tart when he was a psychologist at the University of California, Davis. His subject was a woman named Miss Z who said she had several OBEs per week. Tart took her into the laboratory to investigate the phenomenon.[6] She was able to read a five-digit number that was kept out of her reach or sight. She couldn't have physically gotten up to see the numbers because she was hooked up to an EEG to measure her brainwaves. Any major physical movement would have disturbed the recordings. She said that her consciousness left her body to reach the height necessary to read the number from above. Because Tart knew the number, Miss Z might have learned it through telepathy, but her description is consistent with the OBE type of remote viewing.

Research on remote viewing was done at Stanford Research Institute (SRI) by Russell Targ and Harold Puthoff, two former laser physicists, and Edwin C. May, a former nuclear physicist.[7]

SRI's research was heavily funded by the U.S. government during the Cold War because U.S. intelligence was concerned about the Soviet Union's involvement in psychic research. It would have placed the United States at a significant disadvantage if psychic abilities were real and the United States did not have its own program.[8] After much effort on the part of the researchers, some of the research became declassified in 1995, four years after the Cold War ended.

Most of the research at SRI was conducted with the psychics inside Faraday cages. Because of the hypothesis that psychic abilities might involve electromagnetic transmission, the cages were initially used to see if blocking electromagnetic radiation would impair telepathy and clairvoyance. The shielding actually seemed to help the psychics, which is consistent with later research findings that psychic abilities are greatest when the electromagnetic activity of the earth is lowest.[9] The reason for this is unclear, but a possibility is that the brain's electrical activity is more coherent when there is less ambient electromagnetic noise. And coherency of brain activity, which happens during meditation, appears to be conducive to psychic phenomena.

Many SRI experiments required the remote viewer to draw a picture of what another member of the team was currently observing at a randomly selected distant site. Some experiments used an interviewer to ask the remote viewer questions while he or she concentrated on psychically receiving information about the target. Since an interviewer can unintentionally give subliminal clues when he or she knows the target, the interviewer usually was unaware of it. As another precaution, the interviewer's questions were open-ended and not leading. They were asked the following: *How does the object feel? Is it shiny? What color is it? What do you feel you can do with this object? Does it have an odor?*

Initially the SRI studies were done with Ingo Swann and Pat Price. Swann was a New York artist known for psychic abilities who wrote a book called *Natural ESP.*[10] Price was a retired police commissioner from Burbank, California, who for years had used his psychic abilities to solve crimes.

Price's style was to sit back in his chair and close his eyes. After a brief silence he described what he saw. While working for SRI, Price was asked to help solve the 1974 kidnapping of Patricia Hearst, the nineteen-year-old newspaper heiress who was being held hostage by the Symbionese Liberation Army. He visited the scene where the kidnapping took place and asked to see mug shots of people who had either escaped or been recently released into the local community. Out of forty mug shots, Price correctly identified Donald DeFreeze as one of the kidnappers. He then visualized a scene where DeFreeze had abandoned a car. After he described the car and its location in detail, the car was quickly found.

Swann became bored by the initial SRI experiments, in which he described pictures hidden in envelopes in another room. He successfully convinced the SRI scientists that he could do much more, such as viewing remote locations all over the world without needing an observer at the site. He demonstrated this skill to the intelligence agencies, and the researchers easily obtained funding because this ability was precisely what the military was interested in.

Both Swann and Price became adept at accurately describing distant locations without an observer by being told the location's latitude and longitude or its address. As a further scientific precaution, the geographic coordinates were often presented in binary code, which is a series of 0's and 1's. In one experiment Price sketched a secret Soviet atom bomb laboratory in Semipalatinsk,

Siberia, just from knowing its latitude and longitude.[11] His drawings showed external structures that were previously unknown but later confirmed by satellite photography.

Even more intriguing was Price's description of a Soviet attempt to assemble a sixty-foot-diameter sphere from thick metal sections. Price saw the metal warping in the welding process. Because this construction effort took place inside a secure laboratory building in Semipalatinsk, it could not be verified at the time. But years later the project was described in *Aviation Week* magazine as a 57.8-foot-diameter metal sphere that the Soviets had unsuccessfully attempted to build with thick steel pieces. It was going to be a means of capturing and storing energy from nuclear-driven explosives, but they had the technical difficulties Price described.

The researchers wondered if there was a physical limit to the distance for successful remote viewing. To test this, Swann was asked in 1973 to draw Jupiter just before a NASA Pioneer 10 flyby. He sketched a ring around Jupiter, which was considered an error until NASA discovered Jupiter's ring.

Another result was also considered a mistake at first. Price's drawing of a site in Palo Alto contained some resemblance to the target, but it was less accurate than typical for him. Years later, Targ read an article about the history of that site. The article included a picture that looked just like Price's. The site had been a water processing plant fifty years before the experiment. Structures from the past had been incorporated into Price's picture. The only logical explanation to Targ was that remote viewing isn't limited to present time.

During a demonstration for the CIA, another mistake provided useful information about remote viewing. Although Price and Swann were tested separately, they both described a secret

NSA site rather than the real target, a vacation cabin for a CIA agent that was near the secret site. The NSA site was drawn correctly, but this was still counted as an error because it was not precisely at the provided geological coordinates. Price attributed this error to the impact of secrecy on psychic phenomena: "The more you try to hide something, the more it shines like a beacon in psychic space." Another possibility is that the error arose because of expectation, since the CIA's interest in remote viewing was the possible detection of secret sites. Because the drawings and highly-secret code names were so accurate, a formal investigation was ordered by Congress, which concluded that there was no security breach.

One of the best remote viewers at SRI was Joe McMoneagle. While stationed in Germany he had a near-death experience (NDE) in which he saw his lifeless body on the pavement from a perspective above it. His heightened psychic abilities appeared afterward, which is not an unusual story. Psychics often report that their abilities began after an NDE, which suggests that NDEs can permanently break down barriers to other realms of conscious experience.

McMoneagle has written about his abilities in several books, one of which is *Mind Trek: Exploring Consciousness, Time, and Space Through Remote Viewing.* Another is *Remote Viewing Secrets: A Handbook.*[12] Many of the drawings from his SRI experiments can be found in *Miracles of Mind.*[13] His most amazingly accurate results include the time he drew the locations of a CIA team while the agents were hiding in the San Francisco area. First the agents hid in Lawrence Livermore Laboratory, which is a hundred miles away from SRI. McMoneagle drew many of the laboratory buildings and structures as visible from Lawrence Livermore's West Gate side, including a T-shaped, six-story building that was covered with glass and adjacent to a line of trees. The team traveled to the Livermore Valley Foothills Windmill Farm as their next target, and McMoneagle drew the windmill structures and hills with almost 100 percent accuracy.

Targ and Puthoff's initial experiments lacked control subjects to show the rate of accuracy of people without psychic abilities. Control subjects are usually necessary for psychology experiments to prove that phenomena occur at a greater rate than chance. When they used people with no history of being psychic as controls, many of them displayed psychic abilities.

One control subject was Hella Hammid, a professional photographer who turned out to be one of their most reliable viewers. In one set of experiments she was asked to describe the contents of objects inside aluminum 35 mm film cans. For each experiment, one of ten sealed cans was chosen at random and placed in a park across the street from the SRI laboratory. The use of randomly chosen, sealed cans eliminated the chance that telepathy played a role.

Hammid drew a picture and verbally described what she saw. One can contained a spool of thread and a pin with a prominent head. This was described as "definitely something thin and long, with a nailhead at the end . . . silver colored." Another can had a curled-up leaf. Her drawing had curled lines that she described as "a nautilus shape with a tail." The can with a leather belt key ring led to a picture that looked very similar to the object. Her verbal response was "the strongest image I get is like a belt." Another can was filled with sand, which turned out to be a poor target. The drawing looked like the can itself, which makes sense since the sand took on the can's shape. A gray and white quill inside a can was described as "like a penguin . . . gray and black and white . . . pointed or slightly rounded off at the top . . . open or pointed at the bottom." The results were evaluated by judges who were unaware of the correspondence between the cans and drawings. The judges matched the drawings by their similarity to the can contents, and the number of correct matches was statistically significant.

A total of 411 remote viewing trials were conducted and published over a twenty-five-year period at Princeton University by a psychologist, Brenda Dunne, and an emeritus dean of engineering, Robert Jahn.[14] Like Targ and Puthoff, they found that distance between the target and the viewer didn't matter. However, their success rate declined over the years as the subjects became bored with the experiments and received less feedback on their accuracy. This diminution of accuracy over time has been one reason scientists have been skeptical of the phenomenon. Science demands reproducibility, and scientists expect true effects to either maintain constancy or improve with practice. However, because psychic abilities are more common in creative people, forcing them to perform routine and repetitive tasks in the laboratory might stifle psychic abilities in the same way that it does creativity.

So the most dramatic results are often spontaneous. An illustration of spontaneous remote viewing occurred during a routine ganzfeld experiment by Charles Honorton. This example also demonstrates the power of danger or crisis in evoking psychic skills. The person in the ganzfeld state suddenly became upset and said, "Someone is pointing a gun!" The remark had nothing to do with the target material, but everything became clear soon after the phone rang. It was a security guard alerting them to the fact that a mentally ill person was loose in the halls with a gun.[15]

WHAT THE CLAIRVOYANCE RESEARCH SUGGESTS

Clairvoyance does appear to exist as an ability distinct from telepathy. This is supported by research in which no one knew the target. But in much of that research the remote viewing was based upon geographic coordinates, which makes it difficult for people to believe and relate to. Over a period of two decades, several branches of U.S. intelligence continued to fund the studies, and they were reviewed favorably by the congressional committee for overseeing intelligence. However, because the research was done during the "spy versus spy" mentality of the Cold War and has not been entirely declassified, suspicions have been raised about the true intent of the program. Could the program have been an attempt to persuade the Soviets that we had powerful psychics working for us? That seems unlikely since the SRI research continued for at least four more years after the Cold War ended.

The SRI research has led to several conclusions about remote viewing. For one, accuracy and resolution do not appear to be affected by distance. This is very unusual for any kind of signal processing since electromagnetic signals become weaker with distance. Since the Faraday cages did not interfere, it is even less likely that the mechanism involves electromagnetic waves. So the brains of psychics probably don't receive electromagnetic signals like our televisions and cell phones.

Another conclusion was that study results were less accurate when the psychics knew the target possibilities. As in telepathy, one's preconceptions or expectations can adversely affect results by engaging the brain's analytical capacities. This may be why extremely analytical people appear to have fewer psychic experiences. It may also explain why spontaneous psychic experiences appear to have a higher degree of accuracy than those in the laboratory.

Everything suggests that psychic material becomes available first to the nonanalytical unconscious. Swann and McMoneagle corroborated this when they outlined their process of remote viewing and its distinct stages. Initially they felt kinesthetic sensations and saw fragmentary images. In the second stage they felt basic emotional and aesthetic sensations about the target such as fear, loneliness, or a sense of beauty. In stage three they perceived physical features such as whether the target was large, heavy, or slender. The final stage was when the target's function or purpose became clear. The first two stages involve parts of the brain that are also involved in dreaming: the sensory and emotional processing systems. The last two stages involve analytical parts of the brain that interpret information.

Many SRI "controls" were capable of remote viewing, so another conclusion was that it may be a latent ability in all of us.[16] In order to develop the ability, one needs to learn what the target was afterward, as this feedback teaches people at an unconscious level not to use their imagination. We don't need to know which parts of our brain are activated in order to increase our activation of them by immediate feedback. Psychic abilities are like motor skills and become perfected when the analytical mind isn't in charge of their execution.

Just as biofeedback helps us lower our heart rate or increase the temperature of our hands by providing measurements of those variables, fMRIs could be helpful to developing psychic abilities. Since fMRIs can tell the difference between lying and recalling a true memory, or between the acts of adding numbers and spelling words, fMRIs may help researchers determine if someone is accessing information psychically or just imagining it. This instantaneous feedback could assist the research subject's accuracy in psychic perception.

Chapter 5

THE FUTURE IS NOW: EVIDENCE FOR PRECOGNITION

Predicting the future is easy. It's trying to figure out what is going on now that is hard.

—FRITZ R. S. DRESSLER

PRECOGNITION IS THE KNOWLEDGE of something in advance by some extrasensory means such as clairvoyance. Carl Jung believed in precognition and observed that crisis telepathy could occur before an event, not just during or after it. Alan Vaughn, a coworker of Montague Ullman and Stanley Krippner, kept a diary of his dreams and checked them for evidence of precognition. He was able to identify sixty-one of them as precognitive after the event occurred. For example, on the morning of May 25, 1968, he had two dreams that indicated that Robert Kennedy's life was at risk. Robert Kennedy was shot on June 5, 1968. Was this a synchronicity, or is it possible to look into the future? The evidence suggests the latter may be true.

PETER HURKOS

One of the most famous psychics with precognitive abilities was Peter Hurkos, nicknamed "the scrying Dutchman."[1] He reportedly had no psychic abilities until 1941, when he fell thirty feet to the ground while painting a building. He sustained a skull fracture and was hospitalized in an unconscious state for three days. Immediately after coming to consciousness, he started to receive impressions about people, particularly when they shook his hand. At first both Hurkos and the hospital staff thought that he had gone crazy.

During his hospitalization in Nazi-occupied Holland, he met a man who turned out to be a British secret agent. After their handshake, Hurkos became certain that the agent was about to be killed by the Germans. Hurkos warned a nurse as an attempt to protect the man. After the man was murdered, someone from the Dutch underground came to the hospital to kill Hurkos, because members of the underground assumed that Hurkos's prior knowledge of the murder meant that he was a traitor. Just as the assassin was about to smother Hurkos with a pillow, Hurkos blurted out something in Spanish. This shocked the assassin, because he was thinking in Spanish at the time about how much he hated killing people. He realized that Hurkos had told the truth about having psychic abilities. Hurkos later joined the underground to use these abilities to assist their fight against the Nazis.

Hurkos's particular specialty was called "psychometry," which is the ability to see the past, present, and future by touching related objects. Hurkos described his experience: "I see pictures in my mind like a television screen. When I touch something, I can then tell what I see." In 1956, Hurkos was brought to the United States by Andrija Puharich, M.D., to be tested under tightly controlled conditions at Dr. Puharich's research laboratory in Maine. After two and a half years, Dr. Puharich concluded that Hurkos's psychic abilities were far greater than any he had ever tested.[2] He gave them a 90 percent accuracy. However, as with other famous psychics, Hurkos's abilities became subject to criticism. At times, including when he was tested by the parapsychologist Charles Tart, he was not able to demonstrate them.

PRECOGNITIVE DREAMS

Like telepathy, many anecdotes and research studies on precognition have involved dreams. The following is a famous example:

> About ten days ago, I . . . began to dream. There seemed to be a death-like stillness about me. Then I heard subdued sobs. . . . I arrived at the East Room. . . . Before me was a catafalque, on which rested a corpse wrapped in funeral vestments. Around it were stationed soldiers who were acting as guards; and there was a throng of people, some gazing mournfully upon the corpse, whose face was covered, others weeping pitifully. "Who is dead in the White House?" I demanded. . . . "The President . . . he was killed by an assassin." Then came a loud burst of grief from the crowd, which woke me from my dream. I slept no more that night; and although it was only a dream, I have been strangely annoyed by it ever since.[3]

Abraham Lincoln relayed this dream to his wife and a few friends three days before his assassination. Death by assassination had long been a distinct possibility for Lincoln because of passions inflamed by the Civil War, so his dream might have simply reflected a valid concern. What makes the dream intriguing is its timing and the degree to which it unsettled him. Could he have seen what was forthcoming? Or did he telepathically tune in to the consciousness of the assassin? Or was it synchronicity?

Another example is the following dream by Heinz Pagels:

> I dreamed I was clutching at the face of a rock but it would not hold. Gravel gave way. I grasped for a shrub, but pulled it loose, and in cold terror I fell into the abyss. . . . What I embody, the principle of life, cannot be destroyed. It is written into the cosmic code, the order of the universe. As I continued to fall in the dark void, embraced by the vault of the heavens, I sang to the beauty of the stars and made peace with my darkness.[4]

Pagels was a physicist and mountain climber who wrote about this dream in *The Cosmic Code* six years prior to dying precisely the way he did in his dream. This might have been an anxiety dream of a mountain climber who knew that climbing put him at risk, but unlike most anxiety dreams it bothered him enough to write about it, so it could have been precognitive.

Another category of anecdotal, precognitive dreams involves tragedies of vast impact. Often many people come forth after a tragedy to report that they dreamed about it beforehand. Because these accounts are after the fact, they can only provide weak support for precognition, but they are worth noting. After 9/11, hundreds of people described having had recent dreams in which two planes destroyed the twin towers. The psychiatrist Ian Stevenson described ten cases of precognition of the sinking of the *Titanic*, eight of which involved dreams.

What's more interesting, because it occurred long before the incident, is Morgan Robertson's 1898 novel *Futility*, which was about the wreck of a giant ship named *Titan*. Robertson reportedly created the plot while in a dream-like reverie. The similarities between its details and the *Titanic* disaster fourteen years later are striking. The *Titan* was considered unsinkable, displaced 70,000 tons, was 800 feet long, had 24 lifeboats, carried 3,000 passengers, and sank in April after hitting an iceberg at 25 knots. The *Titanic* displaced 66,000 tons, was 828 feet long, had 24 lifeboats, carried almost 3,000 passengers, and sank in April after hitting an iceberg at 23 knots. Once again, this could be interpreted as either precognition or synchronicity.

Another example was an avalanche of coal waste at Aberfan, a mining town in Wales, which buried 140 people alive in October 1966. So many people came forth afterward with precognitive dreams that the British Premonitions Bureau and the American Central Premonitions Registry were established in 1967.

John William Dunne was a British aeronautical engineer who had a dream on May 7, 1902, that foretold the eruption of Mount Pelée in Martinique. In his dream he warned the French authorities that four thousand lives would be lost. The actual event was in the headlines of his newspaper the next morning. Dunne saw that there were forty thousand deaths, and like Warcollier's fascination with errors in telepathic transmission, Dunne was intrigued that the number of deaths was off by a factor of ten. He hypothesized that he misread the newspaper account of the tragedy in his precognitive dream. Other precognitive dreams led him to believe that dreams can draw from future events as easily as from past events. He developed a theory to account for time travel and published it in 1927 in *An Experiment with Time*.[5] He concluded that the past, present, and future must all coexist, just like in a movie that has already been filmed but is viewed moment by moment.

In *Limitless Mind*, Targ reported that a CIA research associate had a particularly frightening dream about being in a plane crash, which led the associate to delay his flight out of Detroit. He took the dream seriously because of his exposure to research on precognition. The associate's partner still took the flight and died when it crashed. This makes one wonder about the fact that there were many fewer passengers on the four planes involved in 9/11 than normally would have been the case. The Web site 911research.wtc7.net reports that 51 percent of the seats were filled on American flight 11, 31 percent on United flight 175, 20 percent on American flight 77, and 16 percent on United flight 93. According to 911review.org, the average occupancy of flights in the United States was 71 percent. A patient of mine was scheduled to be on one of those planes but for vague reasons decided to stay in Boston another day. Could this have been precognition at an unconscious level? Or just luck? And if it's luck, what is luck? Maybe lucky individuals consciously or unconsciously tap into their psychic abilities.

The CIA research associate's choice to not take the flight leads to a central question raised by precognition: can one favorably change the course of future events after having a precognitive experience? Sometimes this appears to be the case. Here is a dramatic example that was originally printed in J. B. Priestley's *Man and Time*:

> I had some washing to do . . . so I went to the creek . . . I put the baby and the clothes down. I noticed I had forgotten the soap so I started back to the tent. The baby stood near the creek throwing handfuls of pebbles into the water. I got my soap and came back, and my baby was lying face down in the water. I pulled him out but he was dead. I awakened then,

> sobbing and crying. What a wave of joy went over me when I realized that I was safe in bed and that he was alive.[6]

The woman went to a creek the following summer to do washing. As in her dream, she realized that she had forgotten the soap shortly after she had put the baby and clothes down. She went to retrieve the soap, but turned around and recognized a scene right out of her dream. The baby was wearing the same white dress, had picked up a handful of pebbles, and was throwing them in the water. "Instantly my dream flashed into my mind. It was like a moving picture . . . I almost collapsed." Instead of continuing to get the soap, she retrieved her baby and felt that she averted his death by drowning.

RESEARCH ABOUT PRECOGNITION

Although I've provided only anecdotes thus far, precognitive research made its way to the laboratory. Some of the research was done between 1966 and 1972 at the Maimonides Dream Laboratory in New York by Montague Ullman, Stanley Krippner, and Charles Honorton. One of their best subjects was Malcolm Besant, who already had a history of precognitive dreams. The researchers created several dozen "experience cards" that contained descriptions of unusual experiences that the staff would give to Besant. Without showing any of the cards to Besant, they awoke him during REM sleep and recorded his dream reports. The next day, staff members with no knowledge of the dream reports used a random number generator to choose one of the "experience cards," and Besant's dream reports from the previous night were then compared to the experience. In one of these experiments, Besant dreamed of being in a cold white room with small blue objects, while feeling very chilled. The experience card chosen that day instructed the staff to take him into another room, drop ice cubes down his shirt, and blow cold air on him with two blue fans. Of the twelve series of experiments using this protocol, nine had highly significant results.[7]

More than 309 laboratory studies on precognitive experiences while awake were conducted from 1935 to 1987. Sixty-two different investigators used a protocol called "forced choice," which requires the person to guess which one of a fixed number of possible targets would be randomly selected at a future date. Some experiments had subjects choose in advance which of four colored buttons would be illuminated. Others had subjects choose which of five cards would be shown in the future. The time interval between the guess and the target ranged from milliseconds to a year later. The researchers never knew which buttons or cards would be displayed, because they were chosen by a random event generator. Over that fifty-two-year period there were more than two million trials and fifty thousand participants. The results underwent a meta-analysis by Charles Honorton and Diane Ferrari, who found that 37 percent of the studies had positive results, whereas the number of positive results strictly by chance was only 5 percent.[8] Because of the large number of trials, the probability that such results did not occur by chance was greater than a billion billion to one.

The Stanford Research Institute also conducted experiments on precognition. In 1975 they did a series of precognitive trials with Hella Hammid, who was asked to describe where people were going to be at a future time. In one trial she described to Russell Targ a place with "manicured trees and shrubs and formal garden." When he later heard the description of the Stanford University Hospital Gardens from the people in the experiment, it felt like déjà vu because the description so completely matched Hammid's.

Robert Jahn, Brenda Dunne, and Roger Nelson at Princeton University conducted 227 experiments in which a viewer described the future location where a researcher would hide.[9] The accuracy was similar regardless of whether the viewer looked hours, days, or weeks into the future. The odds against the results occurring by chance were 100 billion to 1.

In "associative remote viewing," the viewer is asked to describe something that was arbitrarily paired with an outcome, instead of describing the actual outcome. This is useful when the viewer might struggle with emotional attachment to what happens. It is illustrated by an experiment in which the names of racehorses were randomly paired with household objects, but the viewer was not told the pairings or the objects. Before the race, the viewer was asked to describe the object paired with the horse that would win, and after the race, the viewer was shown the object. Elizabeth Targ, who was a psychiatrist and Russell Targ's daughter, described a "glass sphere" an hour prior to a race in which Shamgo won against great odds. A spherical glass apple juice bottle had been paired with Shamgo.[10]

Dean Radin took a different approach to precognition. He measured the skin conductance of subjects while they watched a computer monitor that randomly displayed images associated with various emotions. Skin conductance changes when we sweat, so it is a good measure of heightened emotions. The images were either calm, such as landscapes, or highly emotional, such as erotic or violent material. The images came from a standardized set of emotional stimuli used in psychology research. Beginning around three seconds prior to seeing the pictures, the subjects' skin conductance changed in a small but statistically significant way. The change was in the direction expected for the picture's emotional content, and was higher when the magnitude of the picture's emotional content was greater. Radin labeled this a "presentiment effect" and conducted four different studies on it, with twenty-four to fifty volunteer subjects each. His overall odds ratio was 125,000 to 1 in favor of a real presentiment effect. One of his subjects was Kary Mullis, a Nobel laureate in chemistry. Mullis found it "spooky" that his skin conductance changed approximately three seconds prior to seeing the pictures.[11]

Another presentiment experiment was conducted by James Spottiswoode and Ed May with 125 volunteers.[12] They used sounds transmitted by headphones instead of images. The participants randomly heard either one second of a very loud sound or one second of silence. There were more fluctuations in skin conductance before the sound stimulus than before the second of silence, which gave the presentiment effect an odds-against chance of 1,250 to 1.

Dick Bierman, a psychologist from the University of Amsterdam, did a fMRI study on ten adult volunteers to see if the presentiment effect showed up in people's brain activity. In women he found a presentiment effect for both erotic and violent images, whereas in men the effect was there only for erotic images. The study was small but suggestive, with odds ratios ranging up to 50 to 1.[13]

PRECOGNITION IN ANIMALS

Like telepathy, precognition may not be limited to humans. Ten percent of "seizure response dogs," trained to look after people having seizures, become "seizure alert dogs." In other words, they are able to foresee a seizure before the person who is going to have the seizure experiences any warning. The dogs can alert them anywhere from seconds to forty-five minutes in advance. How they do so is still a mystery. People often assume that it is due to the dog's acute and highly developed sense of smell, but no scent has been identified with the onset of a seizure. The phenomenon might be akin to precognition.

Another example of possible precognition in animals occurred on the afternoon of August 17, 1959, when thousands of terns, gulls, and other waterfowl flew away after having settled for several months on Montana's Lake Hebgen. Although the lake's surface was virtually without a ripple at the time they left, several earthquakes hit western Montana hours later. When the earth's movements shook the Hebgen Dam, it cracked, and the flooding killed both residents and nearby tourists in Yellowstone. According to the Interior Department's Fish and Wildlife Service, the rangers found no carcasses of any animals in the zone affected by the earthquake. It appeared that animals besides birds were also able to evacuate in time.

There are countless stories like the one above. Is it precognition? Is it an acute sense of hearing? Do animals detect vibrations in the earth or changes in the magnetic field? Dr. J. B. Rhine believed that among the case reports of "unusual behavior in animals there are a fair number of cases in which the reaction is taken to be premonitory."[14] Until studied scientifically, it can't be answered.

At the 1967 winter meeting of the Foundation for Research on the Nature of Man, Robert Morris presented a paper on precognition in laboratory rats.[15] He noticed that when rats were destined to be killed at the end of experiments, they "tend to be more aggressive and recalcitrant" than usual when he took them from their cages. He wondered if they could sense their own death. In order to investigate this he used an "open field," which is an eight-foot-by-eight-foot box whose floor is marked off into small squares. He kept a record of the number of squares the animal typically crossed in a given time. Morris anticipated that rats about to die would show more restricted activity in the open field.

Morris took a group of sixteen rats and ran each of them individually in the open field for two minutes. Immediately after running, each rat was taken to a coworker who either destroyed it or returned it to its own colony, according to a random plan. The open-field scorer did not know which animals were to live or die, and the coworker/executioner did not know about the animal's behavior in the open field. After the series was completed, Morris compared the open-field records of rats who lived with those who died. The half that lived were more exploratory. They left their original square, whereas none of the animals that died did. The results were consistent with the possibility of precognition.

FUTURE MEMORY

Future memory is a very specific type of precognition. P. M. H. Atwater wrote a book titled *Future Memory* wherein she describes it as the ability to "remember the future."[16] The future memory episode is so full of sensory detail that the individual often cannot distinguish it from everyday reality when it happens. It is usually forgotten until it actually happens in the future and something acts to trigger recognition of it. The person then knows what is going to happen next because it is a replay, like watching a familiar movie again.

Atwater did research on more than two hundred cases. The

following quote is from one of them: "I was doing the morning dishes when this rush of energy nearly lifted my head off. I suddenly experienced myself at a dinner party that night, saw who would be there, and took part in what happened and what was said." The woman had no plans for the evening, but she remembered the episode and wondered afterward if she would end up at a party. She then received a phone call that invited her to a party that unfolded exactly as she had experienced it that morning.

Future memory has some resemblance to déjà vu, which is the sense that one has had the experience before. However, future memory episodes are far more detailed than déjà vu and have a characteristic pattern and sequence of events that makes them stand out. Here is a brief summary of Atwater's sequence from beginning to end:

Physical sensation at onset. A rush of heat is coupled with feeling exhilarated.

Present time-space relationships freeze in place. Time stops instantly and the air fills with sparkles.

Expansion. The person feels as if he or she is expanding in size and knowledge.

The future temporarily overlays the present. A detailed experience manifests that feels like a natural component to the person's life, even though it is from the future.

Present time-space relationships resume normal activity. The scenario ends and the sparkles disappear. There is a lingering feeling of pleasure or disconcertedness.

Aftereffects. There is a sensation of being startled or puzzled. The event remains vivid initially but eventually is forgotten.

The future event physically manifests. The memory of having experienced this before is triggered by a "signal," and the person feels as if he or she is acting out a role scripted before. Some feel that they can change the script, while others don't.

Resolution. Experiencing future memory feels uncomfortable at first, but after more experiences the person develops a sense of being more in the flow and a sense of mystery or awe.

Many neurologists would recognize that this description of future memory has many features in common with temporal lobe epilepsy, which has a documented association with déjà vu. Seeing sparkles can occur at the onset of one of these seizures. Distortions in time, space, and body size can also occur in this type of epilepsy, which neurologists have proposed to be the inspiration behind Lewis Carroll's *Through the Looking-Glass, and What Alice Found There.* This does not invalidate the phenomenon of future memory, but suggests that the temporal lobes are involved. Supporting this hypothesis is the fact that nitrous oxide, or laughing gas, can bring on episodes of future memory; and this drug has an effect on structures located within the temporal lobes.

PRECOGNITION AND ITS IMPLICATIONS

There are many reports of precognitive, or prophetic, dreams dating back to ancient times.[17] Precognitive dreams usually feel relevant and pregnant with meaning at the time of the dream, but it is hard to definitively say that these dreams illustrate precognition rather than synchronicity. However, the laboratory research provides support that precognition is possible.

Some support comes from the research on presentiment. Neuroscientists already knew that our unconscious brain perceives events before they reach our conscious awareness, because the first level of processing occurs at the unconscious level. If the content is highly emotional, the unconscious sends signals to the autonomic nervous system, which controls our bodies' visceral responses. This is why we can show changes in sweating, heart rate, and other parameters before we consciously know why. The measurements in the presentiment research show that the unconscious had already started to respond to an event before it occurred. Even though it is only by a few seconds, any ability to perceive the future suggests that we need to rethink our concept of time.

Unlike the other forms of precognition, future memory experiences provide a unique research opportunity because they are very rich in detail and have a characteristic pattern of onset, which makes them easier to recognize when they occur. Also, they can be induced, which makes it possible to do a controlled study of them. One way of inducing them in susceptible individuals is by nitrous oxide, which is also known to cause OBEs. After the induction of a future memory episode, the details could be recorded and compared with subsequent events when they unfold. Although one can't predict when the future events will happen, the events usually last long enough for the subject to alert the researcher when the sequence begins.

Future memory makes the concept of time travel less outrageous. If future memory or any other form of precognition is real, it means the future already exists, which is consistent with what many physicists believe. Einstein's concept that space cannot be separated from time, but rather forms a space-time continuum, makes it easier to accept precognition. If space and time are interconnected and we can see remotely in space, this is consistent with the idea of seeing remotely in time.

Time may appear to us like a forward-moving arrow because of how our brains are constructed to perceive it, not because of how the universe really is. It's adaptive for us to be built this way. Time travel would be very disorienting if it occurred during normal waking consciousness, so this may be why people usually break free of the here and now only in dreams and altered states of consciousness.

If precognition means that there is a preexisting future, it raises certain questions. One is whether we can change the future. Atwater found that some people with future memory couldn't change the future, whereas others could. She told the story of a woman who had a future memory that her husband was in a car accident. She tried to prevent the accident by not letting him go out that evening. However, the accident she envisioned occurred later that week. Another person in her study was able to avert an accident altogether by recognizing the future memory and altering her actions.

Precognition also raises the issue of free will. Our ability to act to change the future is consistent with having free will, but even if the future can't be changed, this does not negate free will. We experience life as a string of moments that are in the present, and it is in the present moment that we can exercise free will. During the filming of a movie, free will is exercised by the director and actors, whose level of performance and improvisations are expressions of free will. But once the film is in the theaters, the same series of events occurs on-screen over and over again.

Another issue has to do with whether precognition is a good thing. If we all have a latent ability to access the future, it might not always be in our best interest to use it. The people who report having accessed information about the future usually only have partial glimpses into it and not the full perspective. This could be a problem if we were to foresee a very difficult time period that we would prefer to avoid. We might become afraid. Foreknowledge also doesn't necessarily mean that we would make the best choices. Some difficult experiences contribute to our personal growth and shouldn't be avoided. Precognition might be real, but it may be accessible to humans only under certain circumstances because that is what is most adaptive.

Chapter 6

MIND OVER MATTER: EVIDENCE FOR PSYCHOKINESIS

The vital force is not enclosed in man, but radiates around him like a luminous sphere, and it may be made to act at a distance. In these semi-material rays the imagination of a man may produce healthy or morbid effects.

—PARACELSUS (1493–1541)

All those who believe in psychokinesis, raise my hand.

—STEVEN WRIGHT, COMEDIAN

PSYCHOKINESIS (PK), OR REMOTE influence, is psychic influence upon an object, a process, or a system. It is the scientific term for practices such as voodoo or witchcraft, healing by prayer, and changing karma by chanting Sanskrit mantras. PK is also associated with more mundane matters, such as the power of positive thinking to bring about desired events. It has become a hot topic in the media lately. *What the #$*! Do We Know!?* and *The Secret* are both popular movies that feature scientists who emphasize that our thoughts are powerful influences on our lives.

Psychokinesis falls into a different category than other psychic abilities. Telepathy, clairvoyance, and precognition all involve the perception of information that is not available through ordinary means. In other words, they are types of extrasensory perception. In contrast, psychokinesis is not a perception but an ability for conscious intention to act as a force on the external world.

When people think of PK they usually think of exotic phenomena, such as the manifestation of material objects out of thin air by gurus in India, the levitation of objects, or other feats that appear more related to magic than anything else. But most of the scientific evidence for PK has been supportive of more subtle phenomena, such as changing the rate at which bacteria multiply in a laboratory dish or the outcome of a coin toss. In other words, conscious intention appears to be more influential in systems already in a state of change or uncertainty, and where one is not attempting to violate fundamental laws of nature, such as gravity.

PK research is divided into three major categories, and the strength of the evidence for PK varies among them. The categories include bio-PK, which was renamed "distant mental interaction with living systems" (DMILS) and refers to the influence on humans, animals, plants, or microorganisms. Macro-PK is the directly observable, large-scale influence upon objects (such as making a table lift up in the air). And micro-PK is influence on random systems or events that requires statistical analysis to be observable.

HEALING BY PRAYER AND SPONTANEOUS REMISSION

One of the most dramatic accounts of healing by prayer is that of St. Peregrine, the patron saint for the spontaneous remission of cancer. More than seven hundred years ago, St. Peregrine was a young priest scheduled to have his leg amputated because of cancer. The night before surgery he prayed with great intensity, and after falling asleep he dreamed that he was cured. Upon awakening, he discovered he had been. His cancer never returned and he lived to be eighty years old, dying in 1345. Such stories are common in the Catholic Church and have been the basis for religions such as Christian Science, which relies upon prayer and faith over the use of medication to cure illness.

Within medicine, spontaneous remission is rare and considered a medical mystery. T. C. Everson and W. H. Cole defined spontaneous remission of cancer as the complete disappearance of a malignant tumor in the absence of all treatment or in the presence of therapy which is considered inadequate to exert a significant influence. They concluded after a review of many cases that spontaneous remission of cancer occurs in 1 out of 100,000 cases. The Institute of Noetic Sciences' book, *Spontaneous Remission*, catalogues several hundred well-documented cases of people who recovered from metastatic cancer even after being close to death, as well as cases of remission from multiple other diseases.[1] Written in 1993, it contains approximately 3,500 references from 800 journals in 20 languages.

There have also been many formal research studies on the efficacy of prayer for cancer and other illnesses. Randolph Byrd, a cardiologist, randomly assigned 393 of the patients admitted to the coronary care unit of San Francisco General Hospital to two groups: one in which they were prayed for and one in which they were not.[2] Those doing the praying were given the first names and diagnoses of patients but were not instructed in how to pray. The patients and researchers did not know whom was being prayed for. The prayed-for patients ended up five times less likely to require antibiotics and three times less likely to develop fluid in their lungs. None of them needed an artificial airway inserted in their throats, whereas twelve of the unprayed-for group required this procedure. Fewer prayed-for patients died, although this was not statistically significant. This is just one of many positive studies on prayer. In fact, enough studies have been positive that Larry Dossey, M.D., author of *Prayer Is Good Medicine,* has publicly stated that doctors who do not engage in prayer for their patients are withholding effective treatment.[3]

However, as the above example of prayer research illustrates, the prayer intentions and outcome measures are often nonspecific, which makes the findings less convincing to mainstream science. Also, the magnitude of the effect of prayer can appear to be less than it actually is because friends and family members often pray for patients in the "unprayed-for" group. It would be unethical and unenforceable to prohibit this outside prayer, so there is not really an "unprayed-for" group, but rather one that received less prayer.

There are a couple of problems in interpreting healing through prayer as a form of psychokinesis. In some studies the patients knew that they were being prayed for, which created the possibility of a placebo effect. And in at least one prayer study such knowledge backfired, because the patients believed that they must be sicker than they thought if people were praying for them. Also, since prayer is a spiritual practice, science cannot eliminate the alternative explanation that positive outcomes are due to divine intervention.

DISTANT MENTAL INFLUENCE EXPERIMENTS ON HUMANS

Some of the earliest research in this category was conducted in the 1920s and 1930s in Russia by Leonid Vasiliev. An English translation of his book *Experiments in Mental Suggestion* was published in 1963.[4] His research was very sophisticated, particularly for the time, and incorporated Faraday cages and other devices to screen out electromagnetic radiation. The distances in his experiments between the subjects and the people doing the mental suggestion ranged from sixty feet to more than a thousand miles. The research team reported that mental suggestion caused changes in breathing, sleeping and awakening states, motor acts, and skin conductance in the distant subjects.

Vasiliev's studies inspired William Braud, Ph.D., of the Mind Science Foundation in San Antonio, Texas. Braud started out as a skeptic but ultimately became a believer and major contributor to the field. His book *Distant Mental Influence* contains many of his published studies, some of which were done with Marilyn Schlitz, Ph.D.[5] Their volunteer subjects ranged from sixteen to sixty-five years old and had answered advertisements in the newspaper and on notices posted throughout the city of San Antonio.

The subject and influencer sat in comfortable chairs, but the subject was in a dimly lit room and hooked up to a device that recorded his or her skin's electrical conductivity. As noted, sweat increases our skin's electrical conductivity, and emotional states can make us sweat, so increased skin conductivity is considered a measure of emotional arousal. The influencer was in another room looking at a readout of the subject's skin conductivity. The influencer used imagery and other techniques, such as changing his or her own emotional state, to try to change the distant subject's skin conductivity. Thirteen experiments that contained ten to forty sessions each were done. The overall success rate was 40 percent, whereas the success rate expected for chance alone was 5 percent.

The results were highly significant, but the participants' comments were the most interesting. One subject reported that he had a very vivid impression of the influencer coming into his room during a session. He felt as though the influencer walked behind his chair and vigorously shook it. The experience seemed so real that it was hard for the subject to believe that it did not really happen. It turned out that the influencer had imagined shaking the subject in that manner as a means of trying to remotely influence him.[6]

EFFECTS OF INTENTION ON NONHUMAN LIVING SYSTEMS

Studies on the effects of intention on single cells don't raise the problem of alternative explanations such as divine intervention, telepathy, or placebo effects. One of the first studies of this kind was done by N. Richmond, who placed single-celled organisms called paramecia under the crosshairs of a microscope. Quadrants of the microscopic field were randomly selected. The experimenter then concentrated on influencing the paramecia to move into the specified quadrant, which they did at a statistically significant frequency.[7] In an experiment by Carroll Blue Nash, former head of the Biology Department at St. Joseph's University in Philadelphia, randomly selected college students used intention to influence bacterial growth, which they successfully increased and slowed down.[8]

In Dr. William Braud's study, thirty-two subjects had a sample of their red blood cells removed and placed in twenty tubes containing water with a low salt content.[9] Dilute saltwater solutions cause water to move into blood cells until they burst. The subjects and tubes were placed in separate rooms and the subjects were asked to psychically protect their cells in ten of the tubes from bursting, while ignoring the other ten tubes containing their cells. The ignored tubes were used as controls, or a basis for comparison with the person's "protected" tubes. The rate at which the red blood cells burst was measured by a spectrophotometer during the fifteen-minute sessions of "psychic protection." The difference between the "protected" cells and the control cells was significant in 9 out of the 32 subjects, whereas only 1.6 subjects would be expected to differ from controls by chance alone. In another experiment, Braud found that people had the greatest influence on their own cells versus those of someone else.

MACRO-PK

Macro-PK is regarded suspiciously even by many parapsychologists because of incidents of fraud. The most famous example was the Israeli psychic Uri Geller, who performed metal-bending feats on television in the 1960s under uncontrolled conditions. Some viewers were so convinced of his power that they reported that some of their household objects were affected. However, many professional magicians accused Geller of using sleight-of-hand techniques, such as surreptitiously replacing spoons with ones previously bent. Also, when tested under controlled conditions he was not able to duplicate his abilities. He later hosted a reality show called *The Successor*, which was a talent search for his psychic heir. On one episode he was caught cheating with a magnet on his fingertip to move a compass needle. This became a huge scandal, especially in Israel, where he was pummeled by the press. Geller claimed that he had good days and bad days, and someone suggested that he felt pressure to cheat on his bad days. Regardless of what the truth is, there is a general distrust among scientists of macro-PK.

In the following experiment, there was no possibility of trickery. Ingo Swann intentionally changed the readings of a device called a SQUID (superconducting quantum interference device), which was buried underneath a building and shielded from electromagnetic influence by several layers of metal. When Swann remotely viewed the interior of the SQUID, he drew accurate pictures of it, but he also changed the tracing of its sinusoidal output, which was normally very constant. The sine wave doubled in frequency when he "projected his consciousness" into the SQUID. His technique appears to be the self-induced out-of-body experience that is discussed in chapter 7. The wave returned to normal frequency when Swann stopped, and it remained fairly constant except when Swann reliably reproduced the change.[10] But even this effect might have been micro-PK.

MICRO-PK

The micro-PK research at Princeton was called PEAR (Princeton Engineering Anomalies Research).[11] PEAR used machines called random number generators, or RNGs. A plot of the generated numbers gives a random distribution curve, which is consistently bell-shaped. Deviation from this bell shape is used in statistics to demonstrate that data are significantly different from chance, which is considered the same as random. The first random event generators were mechanical systems such as tossed coins or thrown dice, but since the 1970s researchers have used devices that rely upon radioactive emission or decay, which is one of the most random processes known to science. That made the emission of individual particles from radioactive sources an excellent target for PK experiments.

Research subjects at PEAR were directed to try to skew the random distribution curves away from the bell shape. During a twelve-year period of nearly 2.5 million trials, 52 percent of the trials showed a shift of the curves in the intended direction, either to the right or to the left. A further analysis found that men were better than women at getting the curve to shift in the intended direction. Women, on the other hand, had a stronger effect on the curve's shape, but not necessarily in the intended direction.[12] Also, the most successful outcomes occurred when experimental subjects looked at archetypal, ritualistic, or religious images while they tried to shift the numbers. These images preferentially activate the right brain, which is associated with the unconscious and intuition. The effect was larger if two unrelated people, rather than one person, tried to influence the curves. If the two people had a close relationship, the results were four times better than for a single person.

This additive effect of people's intention on RNGs led to a series of experiments called the Global Consciousness Project in which the random distribution curves were measured but were not themselves the focus of people's intention. The researchers, Dean Radin and Roger Nelson, were trying to test whether a massive coherency of consciousness could affect the amount of randomness in the world. They hypothesized that this coherency could occur when billions of people focused on the same media event. The researchers saw a dramatic shift in the random distribution curves when the verdict of the O. J. Simpson trial was read on television. A shift also occurred during the opening ceremonies of the 1996 Olympic Games, which were watched by about 3 billion people.

The Global Consciousness Project began in 1998 with three RNG sites. By 2005 the network included sixty-five active RNGs across the globe (throughout Europe, North and South America, India, Fiji, New Zealand, Japan, China, Russia, Africa, Thailand, Australia, Estonia, and Malaysia). The random distribution curves changed during the funeral of Pope John Paul II and around the time of the tragic terrorist acts of September 11, 2001. Hundreds of other high-interest events have been analyzed and found to have significant results.[13]

Radin also did research on the effects of intention in casinos.[14] A baseline for casino profits was easy to find. The percentage of slot machine money kept by casinos is predictable because the ratio is built into the machines. Since gamblers always intend to win, the way to test for PK was to see if gamblers were more successful than the baseline during conditions favorable for psychokinesis.

What conditions are favorable? More than a dozen studies on psychic phenomena have suggested that psychic abilities are greatest when the fluctuations in the earth's geomagnetic field (GMF) are at their lowest.[15] The GMF is in constant flux and is affected by several factors. One is the movement of the earth's molten core. Another is the magnetic fields of other planets when they move relative to the earth. A third factor is the fluctuation in highly charged solar particles that enter our atmosphere. Radin reasoned that if psychokinesis helps people win at casinos, the predictable ratios should shift in favor of the gamblers on days when GMF fluctuations are lowest. He found this to be the case.

In both the gambling and RNG studies, conscious intention was associated with a 2 percent deviation from chance. This small percentage was statistically significant because of the large number of experimental trials involved. Because the effect is small, we would not recognize it without the benefit of statistical analysis. However, this is not the same as saying that a 2 percent difference has a trivial impact. This will become more apparent when I discuss chaos theory, which shows that small changes in initial conditions can lead to a large change in outcome.

Radin suggests that conscious intention is one of the reasons researchers can receive opposite results when studying the same phenomenon. In other words, researchers with a preference for the outcome might influence the results through PK. This may apply to research on psychic abilities. Indeed, skeptical researchers usually have negative results that reinforce their view, whereas believers usually have positive results. One possibility is that skeptics may misinterpret data or set up their experiments less than optimally, but Radin suggests that both believers and skeptics unconsciously influence their data by PK.

PSYCHOKINESIS AND THE MIND-BODY INTERFACE

The influence of the mind over the body is a component of the mind-body interface, which has been of great interest over the past several decades. We know quite a lot more about it now, but in some ways we are not much further ahead than Descartes was in describing a mechanism for the interaction. The following examples illustrate what we know and don't know:

- When we decide that we want a drink of water from a glass in front of us, we move our arm toward it, grab it with our hand, and bring it to our lips. Scientists have outlined the neuromuscular pathway involved. Activity in the brain's motor cortex sends signals to the peripheral nervous system, which ultimately releases chemical messengers at receptors on our arm and hand muscles to make them contract and move. But we still do not know how the desire to drink creates the brain cell activity at the beginning of this physical chain of causality.

- The placebo effect makes people get better solely in response to their belief that they will respond to treatment. It is so well established that pharmaceutical companies are required to prove through clinical trials that their medications are more powerful than a placebo. The nocebo effect causes people who expect side effects from medication to develop them regardless of whether they are on the fake pill. Scientists can discuss the placebo and nocebo effects in terms of the release of various neuropeptides, hormones, cytokines, or other chemical messengers involved in mind-body interactions. But the release of these messengers gets triggered by conscious or unconscious thoughts, and that mechanism is unknown.
- Studies have shown that the visualization of our immune cells attacking cancer actually mobilizes our immune systems to fight cancer. Once again, the mechanism is unknown.
- In a hysterical conversion reaction, someone loses a bodily function rather than face an unpleasant thought. I once saw a woman in the emergency room who had lost the use of her right arm. Her arm's function returned after she was hypnotized and realized that she had an unconscious desire to strangle her husband. Freud was one of the first to realize that our unconscious thoughts can be as powerful as our conscious ones, but how they can temporarily cause loss of function to prevent a thought from becoming conscious is unknown.

- Our thoughts bring about structural changes in our brains. Jeffrey Schwartz, M.D., took PET scans of patients with obsessive-compulsive disorder before and after ten weeks of mindfulness-oriented cognitive-behavioral therapy, a technique that taught them how to redirect their thinking and behavior. The scans demonstrated that their brains changed toward normal brain activity.[16] Other research has shown that when depressed people force themselves to think happier thoughts, they also can "rewire" their brains. Brain scans have shown that psychological treatment can bring about the same changes in activity as pharmaceutical treatment of mild to moderate depression. Over time the pathways involved in normal brain activity are strengthened and new brain cells replace the dead ones. Scientists understand the physiological cascade that causes our genes to turn on and produce the proteins that rebuild brain cells and their connections, but how thoughts can cause the initial step in this chain of events is unknown.
- Conscious intention can also act on something artificially attached to our body. At the University of Pittsburgh, Andrew Schwartz electronically linked a robotic arm to a monkey's brain.[17] The monkey's arms were left intact but restrained so that the monkey could only use the robotic arm to grab food and bring it to its mouth. Ninety-six electrodes, each thinner than a human hair, were attached to the monkey's motor cortex in the area for controlling muscular movement of one of the restrained arms. The exact placement of the electrodes was not critical, and Schwartz used far fewer brain cells to supply information to the artificial arm than the brain

> normally uses to control an arm. The monkey successfully learned how to manipulate the robotic arm by intention in order to eat. So even when the connections are artificially created, the desire to eat can generate electrical impulses in the necessary brain cells, although precisely how this happens is still unknown.

All of the above examples illustrate that our thoughts stimulate activity inside our bodies. PK suggests that they can also act as stimuli outside of our bodies. This goes against the mainstream model of consciousness, which assumes, without proof, that our consciousness and thoughts are confined to our brains. In fact, the psychic research already presented, and the modern physics research presented in chapter 10, suggest the opposite.

PK brings us back to the question of what consciousness is and how it interacts with the physical world. According to mentalism, our perception of the physical world is an illusion and everything is a product of our mind, so the mind's ability to influence the physical world is simply a function of one's state of consciousness. In dualism, consciousness and the material world are totally separate and distinct, so the way they interact with each other is mysterious whether it happens locally or from a distance.

In the material model, everything, including thoughts, can be reduced to something material. That model could account for PK if it considered consciousness to be a force that could act locally and at a distance. There is a precedent for forces like this. Gravity is a force that acts at a distance, and although we live with it constantly and have theories about how it works, it is still a mystery.

So regardless of one's model for consciousness, PK is similar to the interaction between our thoughts and bodies, but it can act remotely on other systems or things. If one applies Occam's razor and doesn't evoke a more complicated explanation when a simpler one will do, we need to discard the assumption that the interaction between mind and matter differs with distance.

WHAT CAN WE CONCLUDE FROM PSYCHOKINESIS RESEARCH?

There is evidence for PK, but it differs among the various types. Although many positive research studies exist for prayer's effect on healing, this effect is not clearly PK, because of the possibility of divine intervention. Macro-PK research has been tainted by fraud, but there is at least one case that demonstrates it. The data on micro-PK, such as that on the random number generators, show a small but significant effect. And the research on single-celled organisms, human red blood cells, and skin conductivity of humans demonstrates significant effects.

PK may be an extension of the impact our thoughts have on our own bodies. Consistent with this is Braud's research on living systems, which shows that we have the greatest PK impact on systems most closely related to our own. We influence gerbils more than fish, people more than gerbils, and our own cells more than those of someone else.

When people focus on the Olympics and don't even know that RNGs exist, changes to the RNG curves are due to a kind of effect that is different from what occurs when someone is asked to focus on skewing the RNG curve. It suggests that a collective level of consciousness can affect the world and make it less random. It would have to occur at the subatomic level because the random number generators operate by radioactive emission of subatomic particles. We know from research in quantum physics that the subatomic world is affected by our observation or consciousness, so this is not as far-fetched as it may seem.

The Global Consciousness Project suggests that focused consciousness can have an ordering effect upon the universe, which otherwise has a tendency toward disorder or entropy. Living systems have a self-organizing effect on the atoms that comprise them, whereas entropy takes over as the body decays. The Global Consciousness Project (along with PK, synchronicities, and telepathy) also suggests that we live in an interconnected universe. This interconnectivity has been proposed by mystics for millennia, and is now validated by modern physics.

Chapter 7

WAS SHE OUT OF HER MIND OR JUST OUT OF HER BODY?

> *Our normal waking consciousness, rational consciousness as we call it, is but one special type of consciousness, whilst all about it, parted from it by the filmiest of screens, there lie potential forms of consciousness entirely different.*
>
> —WILLIAM JAMES

MANY, BUT NOT ALL, psychics report that their consciousness is outside their bodies when they engage in psychic tasks like remote viewing. William James's comment in the epigraph to this chapter was inspired by the out-of-body experiences he had while under the influence of laughing gas. James's conclusion that the brain is merely a filterer of consciousness, rather than what creates it, is shared by several members of the faculty in psychiatric medicine at the University of Virginia. They provide extensive arguments and research in support of this perspective in their 2007 book *Irreducible Mind.*[1]

This is actually an ancient view. Eastern philosophies teach that we live in a sea of unified consciousness; that our personal consciousness exists both in and outside our bodies; and that our consciousness can travel away from the body during dreams, out-of-body experiences (OBEs), and at death. In other words, the brain is not the generator of consciousness.

The Eastern view of consciousness is radically different from the mainstream brain-based model. The Eastern model persists for many reasons, most of which come from experience and not just beliefs. Many authors have attempted to validate the Eastern model scientifically by pointing out the parallels between mystics' experience of consciousness and quantum physics, both of which describe an interconnected universe with more dimensions than we experience through our senses. But even if the brain doesn't generate consciousness, it has a role in psychic phenomena. For the most part, what happens in the brain has not been discussed in literature about psychic experiences, but we now have some important clues.

OUT-OF-BODY EXPERIENCES

Ingo Swann and Edgar Cayce were among those who used self-induced OBEs to obtain psychic information. During an OBE people report that their "self," or center of awareness, is located outside of the physical body. They have the sensation of floating, traveling to distant locations, and observing the physical body from a distance.

Other than those shared features, the experience is highly variable. Only 7 percent report seeing an astral cord, which is described as a silver elastic cord that connects their physical body with their traveling "astral" body. In Eastern philosophy the astral body is one of four major "subtle bodies," which are invisible sheaths or layers of a consciousness field that flow through and around the physical body. Only the astral body engages in distant travel, which is why self-induced OBEs are also called astral projection. Some people "see stars" during an OBE and interpret that as traveling in outer space, so *astral* was chosen as this subtle body's name because it is derived from the root word for "star." Many experience the astral body during their OBE as a more translucent version of their physical one. But not everyone experiences an astral body; some people report that they are "pure consciousness" or "balls of light" during their OBEs.

OBEs also don't require being in a trance, as Edgar Cayce was during his readings. Sometimes the person can be awake and simultaneously maintain an in-body perspective. For example, a twenty-two-year-old patient of mine told me about his intermittent out-of-body experiences. "I would be walking down the street, then suddenly I was observing the entire scene from up above while at the same time I was seeing and interacting with everything from inside my body. It freaked me out at first, but after a while I got used to it." A previous psychiatrist had diagnosed him as having schizophrenia and started him on an antipsychotic medication that stopped his OBEs. They returned when he discontinued the medication. Although antipsychotics controlled his OBEs, this does not mean that OBEs should be written off as hallucinatory experiences. What we can conclude is that antipsychotics can restore the usual in-body perspective.

OBEs differ from hallucinations in that they are often verified. An average of 19 percent of people who have OBEs report that they verified what they witnessed during them.[2] One example comes from Miss Z, the woman who read the remote five-digit number while in an OBE at Charles Tart's laboratory. When she was fourteen she had an OBE in which she witnessed the rape and stabbing of a girl wearing a checkered skirt. The crime occurred in a particular section of her hometown that she recognized in her OBE. The next day the newspaper reported that a girl in a checkered skirt had been raped and murdered that night, in the precise location and in the same way as it had occurred in Miss Z's OBE. This eerie parallel had a traumatic psychological impact on Miss Z, who became frightened of having more of those experiences.

Carlos Alvarado, an expert on OBEs, independently verified the reports of distant events seen during OBEs in 5 percent of his sixty-one cases.[3] Although this is not a high percentage, it is sufficient to suggest that some people do see events from a perspective impossible from their physical body.

Although the concept of an OBE strikes many as bizarre and incredible, reports of out-of-body experiences are actually not uncommon. Based upon five studies, Alvarado reported that 10 percent of the general population has experienced one or more. A combined analysis of forty-nine studies of students showed that they have an even higher prevalence of OBEs, 25 percent, which is two and a half times that of the general population.[4] The increased percentage of reports in students may be due to several factors. Students may have more open-mindedness about the experience, greater willingness to report it, and a higher desire to induce an OBE. They do have a higher incidence than older generations of using recreational drugs that induce OBEs, such as ketamine, DMT, and nitrous oxide (laughing gas). The increase could also be a manifestation of the brain's continuing evolution. If so, it would appear to a greater extent in the youngest generations.

There are several ways in which an OBE can be induced besides drugs. OBEs can be self-induced by meditation, self-hypnosis, or other psychological techniques for astral projection. Illness, stress, or pain can cause an OBE. I've heard accounts of people who had OBEs after serious accidents, in which the OBE removed them from feeling unbearable physical pain. OBEs can also occur as an escape from severe emotional pain, especially if the emotional pain has its origins in childhood trauma. And spontaneous OBEs can occur without induction by any of the above factors, particularly after one has had several previous OBEs.

The majority of the scientific community regards OBEs as a type of illusion. Dr. Henrik Ehrsson, at the University College London Institute of Neurology, investigated this idea by creating the illusion of an OBE in normal subjects. Cameras were placed a little over two feet behind their heads. The subjects wore head-mounted video displays that showed them the live film recordings from these cameras. Each eye received information from a different camera to create a stereoscopic (3-D) effect. The created visual effect was as if they were watching themselves from behind. The researcher also used a plastic rod to touch the subjects' actual chests after they had seen another plastic rod disappear from the camera's view in the direction of the illusory body's chest. The tactile input added to their sensation that they were seeing their body from behind. Sensors measured their perspiration as a marker of their anxiety when the cameras showed a hammer approaching their illusory chests in a threatening way. They perspired more when they saw the approaching hammer, so to some degree they believed that the illusory body was their body.[5]

Ehrsson's research was cited in the media as an explanation for OBEs. However, he created the *illusion* of an OBE by artificially providing a view from the back of subjects' heads; he didn't induce one. Other experiments have shown that when the brain receives visual and tactile input that are in conflict with its input for body location, the visual and tactile input override the body's signals.[6] This artificial OBE is only another example of how the brain can be fooled by the way it prioritizes sensory input. It still leaves the mystery of how accurate alternative views are created during OBEs.

These accurate accounts of distant locations is one reason the Eastern view persists, but another reason is that people have reported that they sensed the presence of a distant friend or relative when that person came to them in an OBE. A study related to this was done by Dr. Robert Morris at the Psychical Research Foundation in Durham, North Carolina. He worked with Keith Harary, who could reliably have an OBE at will. Harary's rambunctious pet kitten was used as a test subject, since the kitten had a history of quieting down only in Harary's presence or when Harary astral-projected to be there. Morris found a statistically significant correlation between the periods when Harary astral-projected and when the kitten was quiet.[7] The results were intriguing, although on subsequent testing the effect decreased. The results may be evidence for astral projection, but an alternative explanation is that the kitten quieted down because of telepathic communication. Either way, they suggest a link between the consciousness of Harary and his kitten.

THE NEUROLOGY OF OBEs

One of the first studies of the changes in brains during OBEs was done by Charles Tart, who made EEG recordings of Miss Z on four nights in his laboratory.[8] She was instructed to awaken after an OBE so that he could note the time and look at the preceding recordings for correlations between her OBEs and her brainwave activity. He found that her sleep EEG was very unusual in general, and the recordings associated with her OBEs differed from those she had during normal REM sleep. During the OBEs she had a poorly developed EEG pattern with some elements of dreaming sleep, but it lacked the rapid eye movement that is almost always associated with dreams. The EEG during OBEs was also mixed with transient periods of wakefulness. William Dement, a renowned sleep expert, agreed with Tart that her EEG during OBEs was not a recognizable pattern.

A subsequent study by Michael Persinger and colleagues included both MRI imaging and EEG recordings of Ingo Swann when he was remote viewing.[9] Unlike Miss Z, who was sleeping during her OBEs, Swann had OBEs while awake and drawing remote images. The research team consistently found a distinctive and unusual EEG pattern, not typical for normal vision, over the occipital lobes when he was remote viewing.[10] The occipital lobes are where the visual cortex is located, so brain activity there supports the idea that Swann was visualizing something during remote viewing.

The researchers suspected that the pattern related to ponto-geniculo-occipital (PGO) transients because of its distinctive waveform. The name *ponto-geniculo-occipital* refers to the pons (a region of the brainstem associated with sleep), the lateral geniculate body (a section of the thalamus that processes visual information from the eyes), and the occipital lobe (which contains the visual cortex that receives visual input from the lateral geniculate body). The brain interprets PGO activity as sensory information and uses it to direct and determine dream content. The researchers concluded that if the unusual EEG activity was PGO, then "the images reported by Mr. Swann would be analogous to a special variant of waking dreams."

Unidentifiable "bright objects" were found in Swann's brain

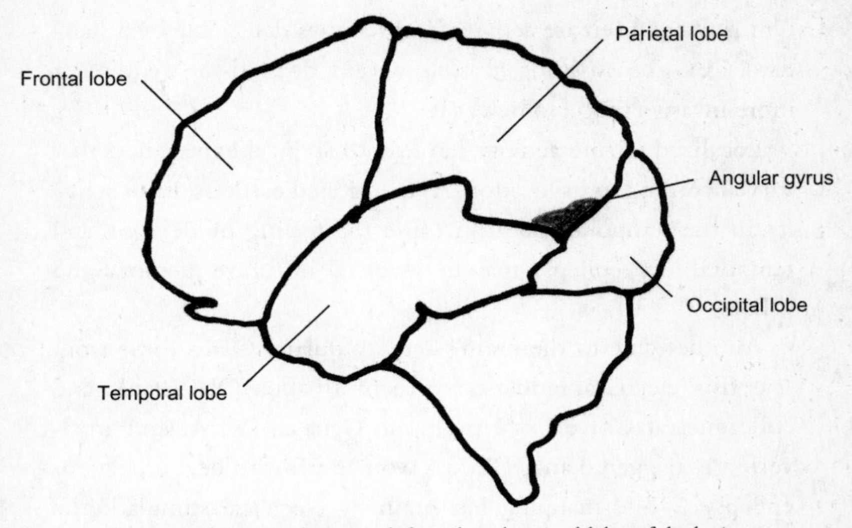

Figure 2. The occipital, temporal, frontal, and parietal lobes of the brain.

by MRI. These objects were probably deposits of a material such as calcium that provides a brighter MRI signal than normal tissue and accumulates in damaged tissue. The objects were mainly located where the parietal and occipital lobes meet, and along the pathways where they send and receive information. The parietal lobes assist in the orientation of our bodies in space and with visual-spatial functions. Their sensory input enables the parietal lobes to create topographic maps of our physical surroundings, which in turn let us imagine what objects would look like from different angles.

Aberrant brain tissue, such as Swann's "bright objects," can cause localized abnormal electrical activity called epileptic or seizure activity. A normal EEG pattern reflects sequential firing of cells along discretely defined pathways. But when adjacent brain cells fire inappropriately in sync with each other during a seizure, it appears on the EEG as large atypical spikes or waves.

Unfortunately, seizure activity is not always detectable by a standard EEG, because localization within deep tissue requires a more invasive probe to detect it.

Localized seizure activity can lead to atypical experiences that vary according to its location. As mentioned earlier, seizure activity in the temporal lobe can cause the feeling of déjà vu, and temporal lobe epilepsy may be involved in future memory and OBEs.

Another clue to the brain's activity during OBEs came from scientists' electrical induction of them. In 2002, Olaf Blanke and colleagues at University Hospital in Geneva, Switzerland, inadvertently triggered an OBE in a woman prior to her surgery for epilepsy. While mapping her brain by electrical stimulation, a stimulus at full intensity in one particular region led the patient to report, "I see myself lying in bed, from above, but I only see my legs and lower trunk."[11] The region they stimulated was the right angular gyrus, which is where the temporal and parietal lobes meet in the right half of the brain.[12]

A subsequent study by Blanke and colleagues used MRI brain imaging to study patients with a high incidence of OBEs.[13] These patients invariably had damage to either the left or right angular gyrus. The researchers also inserted electrodes into these patients' brains and discovered seizure activity in the region of the angular gyrus.

Blanke's findings reminded me of an MRI study in which the angular gyrus was larger on the right than on the left in schizophrenia, which is the reverse of what is normal.[14] Since 42 percent of people with schizophrenia report OBEs, an abnormality of the angular gyrus may be the reason schizophrenia and OBEs commonly occur together.[15] An abnormality of the angular gyrus might also apply to Miss Z, since Dr. Tart's paper reported that she might have schizophrenia. But even if Miss Z had schizophrenia, that does not negate the fact that she was able to see remote numbers during her OBEs.

The angular gyrus is strategically located at the crossroads of areas specialized for processing touch, hearing, and vision. Blanke proposed that abnormal electrical activity in such a rich sensory area might cause the brain's sensory signals to get mixed up and thus lead to misperception of where one's body is located. It would create an illusion like the one Ehrsson artificially created with his cameras. But the problem with this argument is that it doesn't address how the brain could correctly access remote information unavailable to the usual senses. I propose that unusual electrical activity in the angular gyrus and its adjacent temporal lobe can lead to accessing psychic information via an OBE.

The primary function of the angular gyrus is the symbolic representation of the world. This is probably why it is proportionately much larger in humans than in other primates, and why it is usually larger on the left side of the brain, because most of us are right-handed and the left angular gyrus is where most right-handed people perform most language functions. Left-handed people can have language functions on the left or right side, as well as divided between the two. Damage to the angular gyrus can impair our ability to understand language, perform numerical calculations, and understand metaphors. The angular gyrus also assists in dreaming, which is rich in symbolism and metaphor.

My own proposed model of the brain combines the following facts: the angular gyrus is associated with both OBEs and dreams; Swann's OBEs were thought to be a form of waking dreams; both dreams and OBEs have been credited with supplying psychic information; and many people have occasional dreams in which they can fly, thereby traveling as one does during OBEs.

These all show an overlap between OBEs and dreams. Therefore, heightened activity in the angular gyrus may cause OBEs, which may be "waking dreams."

That OBEs are related to dreaming has been suggested before by Stephen LaBerge, a noted authority on lucid dreaming who believes that astral projection is a variant of lucid dreaming.[16] In lucid dreaming, one's conscious mind wakes up while one is still dreaming, recognizes the experience as a dream, and can take control over what happens next in the dream. Most people have a lucid dream at least once in their lives, but 20 percent of people in the United States report having lucid dreams one or more times per month. Astral projection entails the reverse sequence of events from lucid dreaming, but with a similar result. It requires that one keep the conscious mind awake while otherwise going into dreaming sleep.

Kevin Nelson, Ph.D., and colleagues at the University of Kentucky interviewed fifty-five people who had near-death experiences and found that many reported a prior history of OBEs when transitioning to sleep. The OBEs were very similar to the ones they had during their NDEs. Because the OBEs occurred during transition to sleep, Nelson concluded that OBEs are a blending of wakefulness and REM sleep.[17]

To have a lucid dream, one starts dreaming and uses a preselected trigger to signal that it is a dream. Carlos Castaneda talked about lucid dreaming in his books. He told himself that he would know he was dreaming when he observed himself looking at his hands. There are also devices activated by the eye movements of REM sleep that can be worn during sleep to promote lucid dreaming. The devices send light signals through the eyelids that can be perceived without waking the dreamer.

OBEs can be a variant of dreaming even when the person's

body doesn't appear to be asleep or in a trance. Swann did not appear to be asleep during his OBEs, but he had brainwaves consistent with "waking dreams." This can occur because the components of the dream cycle can become desynchronized from each other and occur separately. For example, the body normally becomes paralyzed during dreams so that we don't act out the dream, but the paralysis usually stops immediately upon awakening. There is a medical condition called sleep paralysis in which the person awakens, but all of the body (except for the eyes) remains completely paralyzed. This can continue until something snaps them out of it, like the sound of an alarm clock or a nudge from a spouse. During sleep paralysis one can experience hypnopompic hallucinations, which occur during the transitional state of consciousness when one is waking up, and hypnagogic hallucinations, which occur when one is falling asleep. The hallucinations come from unconscious dream material that is projected by our brain onto the background scenery of where we are transitioning to or from sleep.[18]

Miss Z had frequent hypnagogic hallucinations, which isn't surprising given her aberrant dream cycle. One connection between these hallucinations and OBEs is that some people enter into OBEs by focusing on these hallucinations. It helps them to maintain conscious awareness while their bodies fall asleep, but it also is a sign that they have a disrupted sleep-wake cycle.

There are other examples in which the dream cycle's components can fall out of sync with each other. Sleepwalking and sleep talking are the reverse of sleep paralysis. The person is dreaming but does not have the body paralysis to prevent him or her from acting out dreams. Given the different ways in which the sleep cycle can become disconnected, and evidence from the EEGs of people who have had OBEs, a reasonable hypothesis is that OBEs can provide dream-like access to information that is normally unconscious without the other components of sleep.

NEAR-DEATH EXPERIENCES

> *He's not dead, he's electroencephalographically challenged.*
>
> —Anonymous

In 1982, the pediatrician Melvin Morse reported a case which he called a "fascinoma." He resuscitated Crystal, an eight-year-old girl who had drowned in a swimming pool. She had been without a heartbeat for nineteen minutes when he arrived on the scene. Her pupils were fixed and dilated, which is one sign of coma or brain death. Dr. Morse got her heart beating again, and she regained consciousness three days later. After a few weeks, Dr. Morse ran into her at the hospital. Crystal said to her mother, "That's the guy who put the tube in my nose at the swimming pool." Morse was amazed she could remember something that happened while she was brain-dead. Crystal also described a classic near-death experience, including a trip to heaven and being told to return to help with her yet-to-be-born baby brother. Morse published the case in the *American Journal of Diseases of Children*.

NDEs usually refer to an event in which someone is pronounced clinically dead but is resuscitated. The person typically has no pulse or breath for an average of ten to fifteen minutes, but the state can last up to an hour. Remarkably, many people don't suffer brain damage after a NDE with medical circumstances that normally cause severe loss of cognitive function. In fact, some reports are of enhanced cognitive or psychic abilities afterward.[19]

The term "near-death experience" wasn't coined until the 1970s, but these experiences have been written about for thousands of years. Plato's description in *The Republic* involved a soldier named Er who was considered dead and placed on a funeral pyre but who then revived and talked about his experience. NDEs were once thought to be rare, but Kenneth Ring estimated that at least one third of people who come close to death report them.[20] In 1997 *U.S. News and World Report* estimated that 15 million Americans have had one.

Raymond Moody, M.D., was a pioneer in NDE research. His book *Life After Life* was published in 1975 and described fifteen common features of NDEs: ineffability, or the inability to describe the experience in words; hearing oneself pronounced dead; feelings of peace and quiet; hearing unusual noises; seeing a dark tunnel; finding oneself outside one's body; meeting spiritual beings; a very bright light experienced as a "being of light"; a panoramic life review; sensing a border or limit to where one can go; coming back into one's body; frustrating attempts to tell others about what happened; subtle "broadening and deepening" of life afterward; elimination of the fear of death; and corroboration of events witnessed while out of the body. Two years later he added four commonly reported experiences: the existence of a realm where all knowledge exists; visiting cities of light; a realm where bewildered spirits exist; and supernatural rescues, such as by angels.

The definition of NDEs has broadened to include people who have had similar experiences who didn't actually die and come back to life. A review of medical records revealed that slightly more than half the patients who report NDEs were never in danger of dying, even though they were ill enough to have been hospitalized.[21] Owens and colleagues also found that certain features of the NDE, such as an encounter with a brilliant light, enhanced cognitive function, and positive emotions afterward, are more common in those individuals who were actually closer to death than those who were not as seriously ill.

After they were brought back to life, near-death survivors typically can accurately describe what happened while they were clinically dead. They remember what people said, what equipment and drugs were used, which people were present, and how people were dressed during the resuscitation. Frequently the details of what they saw during the NDE could only be observed from a vantage point outside the body. For example, one unconscious patient accurately told hospital staff about a tennis shoe that was on the ledge of the hospital's third-story window. Another identified the nurse who had removed his dentures and the drawer she had placed them in while he was in a coma. More remarkably, Ring and Cooper reported thirty-one cases of blind individuals who described objects and events during their NDE that were visually accurate. This was even reported for those blind from birth.[22]

One theory about NDEs presented by neuroscientists is that the phenomena associated with NDEs (the tunnel of light, visions of angels, etc.) are a result of lack of oxygen to the brain. They cite the fact that fighter pilots have experiences that share many features of a NDE when they undergo brief periods of unconsciousness because of rapid acceleration. The pilots have a sense of floating, dissociation from their body, and pleasurable sensations. OBEs are not common for fighter pilots but have been reported when the blood flow to their brains was compromised. The pilots also have reported tunnel vision, bright lights, and visiting beautiful places that include family members and close friends. However, they haven't reported some other NDE features such as a life review or panoramic memory of their life.[23]

One problem with applying the oxygen deprivation theory to NDEs is that patients who suffer from insufficient oxygen to the brain report NDEs less frequently than people whose brush with death didn't involve oxygen deprivation. Also, prolonged oxygen deprivation usually causes severe brain damage, but there is no evidence for this in survivors of NDEs. In fact, the IQs of several children actually increased after their near-death experiences.[24] And one study measured blood levels of oxygen and carbon dioxide during NDEs and found they weren't correlated with the experience.[25] However, the fighter pilot data still suggest that brief oxygen deprivation may be a trigger factor for an OBE with some features of NDEs.

Skeptics criticize many reports of NDEs because they did not occur under carefully controlled conditions. A case described by Dr. Michael Sabom avoids that criticism because it was set up as an experiment on NDEs. The subject was a woman who required brain surgery for a large basilar aneurysm, which was a dilated blood vessel at risk for bursting at the base of her brain. The operation required that her body temperature be lowered to 60 degrees Fahrenheit. Her heartbeat and breathing ceased, her brain waves flattened on the EEG and met the criteria for brain death, and blood was completely drained from her head before the aneurysm was repaired. She was under extremely close observation during this critical time.

After she was brought back to consciousness, she reported the classic near-death experience of the tunnel vortex, bright light, and deceased family members coming to greet her. But she was also able to report specific details of what was said and done by medical staff while she was in the brain-dead state and had a flatline EEG.[26] These details helped to pinpoint the time when she saw the tunnel and deceased relatives as the same time she was brain-dead. This provided evidence against the argument of skeptics that NDEs occur right before people lose consciousness. Equally important, the NDE events occurred while her eyes were taped shut and her ears had molded earplugs, to prove that the patient's experiences weren't occurring through normal sensory routes.

Michael Persinger has tried to simulate NDEs through the application of complex magnetic fields, which can induce electrical activity in the brain at specifically targeted areas. Since NDEs have a strong spiritual component, and substantial research has connected the temporal lobe to spiritual experiences, he chose to target the temporal lobe. Also, that lobe contains the hippocampus, a brain structure highly sensitive to the oxygen deprivation some scientists consider to be a cause of NDEs. When he induced activity in their temporal lobes, some subjects had out-of-body experiences, floating sensations, and feelings of being pulled toward a light. They reported hearing strange music and feeling that they were experiencing something with profound meaning. But the induced experiences also differed from NDEs. The subjects felt dizzy, which doesn't happen during a NDE, and most of the experiences contained only fragments of the NDEs outlined by Moody.[27]

Persinger had altered the brain activity in a way significantly different from how it is altered when a person actually dies and is brought back to life. He experimentally increased activity in the temporal lobes, whereas the EEG during death and in the case Sabom discusses shows no obvious electrical activity in the brain. But Persinger didn't think that his data provided a full explanation of NDEs or that it challenged the validity of subjects' perceptions during NDEs. He concluded that the brain's state during NDEs may simply be more conducive to perceiving otherwise inaccessible information, including realms that contain dead relatives and angels.

Persinger's data contribute to the evidence that the temporal lobes are involved in out-of-body experiences, including NDEs. One piece of evidence was the Swiss study that brought on a partial OBE by the stimulation of the angular gyrus. The Swiss study only induced a minor version of an OBE, so the whole experience probably requires the temporal lobe to become activated along with the angular gyrus. In addition, Willoughby Britton, a graduate student in psychology at the University of Arizona, found that 22 percent of the people with a history of NDEs exhibited the electrical pattern in their EEGs associated with epilepsy or seizures in the left temporal lobe. This change in electrical activity suggests that their brains sustained mild permanent damage to the temporal lobe during the NDE. The lack of preexisting EEGs makes the data less conclusive. However, it is consistent with the hypothesis that temporal lobe epilepsy can lead to OBEs and psychic phenomena, both of which increase in frequency after an NDE.

Evidence for post-NDE changes to the brain's dream circuitry comes from Britton's sleep studies on twenty-three people with a history of NDEs and twenty-three matched controls without any history of a life-altering event. In people who had a history of a NDE, REM (rapid eye movement or dreaming) sleep began 110 minutes after the onset of sleep, as opposed to the normal 90 minutes. In other words, NDEs appear to change the brain so that the normal stages of sleep are disrupted.

Another important brain region implicated in NDEs is the limbic system, which is partially contained inside the temporal lobes. Among other functions, the limbic system is involved in forming social bonds, laying down long-term memory, assigning what emotions to feel in response to people and things, and REM sleep. Activation of the limbic system by strong painful emotions can cause OBEs. Drugs that cause OBEs, such as ketamine, also act on the limbic system, which is rich in receptors for glutamate. These are called NMDA receptors and are where the anesthetic ketamine binds. Ketamine can also cause people to see tunnels, lights, and mystical entities with whom they believe they have telepathic contact.[28] PCP and nitrous oxide (laughing gas) can have a similar effect. These drugs are all called dissociatives, because they dissociate the mind from the body.[29]

K. Jansen proposed that the NMDA receptor is involved in NDEs and that there is a ketamine-like substance released in the body under the conditions that produce them.[30] Jansen acknowledged that identifying the possible neurochemicals involved in NDEs does not explain away the phenomenon but may be a key to the "door to a place we cannot normally get to; it is definitely not evidence that such a place does not exist."

N,N-DIMETHYLTRYPTAMINE (DMT)

The schizophrenic is drowning in the same waters in which the mystic swims with delight.

—Joseph Campbell

DMT is an extremely short-acting and powerful psychedelic found in many plants. It was first discovered in human blood in 1965 by a German research team, and in 1972 the Nobel Prize–winning scientist Julius Axelrod found it in the human brain. It can explain the brain's chemical induction of NDEs and OBEs.

At first, scientists were intrigued with the possibility that excessive DMT may be the cause of schizophrenia, but interest in this hypothesis died down even though it was not disproved.[31] One problem with the research was that scientists were looking at levels of DMT in the blood, which are a poor reflection of the amount of DMT inside the brain, and it is the levels of DMT in the brain that are significant. In addition, a moratorium on psychedelic research had been imposed in 1976 by the U.S. National Institute of Mental Health.

Rick Strassman, M.D., is a psychiatrist who has conducted clinical research on DMT at the University of New Mexico.[32] After many bureaucratic delays, Strassman was able to get permission in 1990 from the Drug Enforcement Agency and the Food and Drug Administration for his university to conduct research on the effects of injecting DMT intravenously into volunteer human subjects. He chose people who had a prior experience of DMT because they were the most capable of giving well-informed consent.

Some of the volunteers' prior exposure was from ingestion of ayahuasca, a beverage made from a plant with DMT as its active ingredient. Ayahuasca is believed to have inspired Native American rock art with psychedelic qualities and has a long history of being consumed by Peruvian shamans to induce an altered state of consciousness. During that state they have dream-like visions that reportedly provide cures for their patients. For the past 150 years, botanists, anthropologists, and others have visited Amazonian shamans and ingested ayahuasca as part of their own research into its effects. They too have reported seeing and hearing visions, often in the form of serpents speaking to them about the potential healing power of the indigenous plants.

After receiving an injection of DMT, some of Strassman's

subjects said, "I no longer had a body," "My body dissolved—I was pure awareness." Many saw extremely bright colors and so many overlapping images that they couldn't distinguish the background from the foreground. This gave them a sense of being "beyond dimensionality." They also saw specific visions such as "the tree of life and knowledge," "tunnels," "stairways," "the inner workings of machines or bodies," "reptiles," and "DNA double helices."

Strassman considered doing brain imaging studies of his subjects under the influence of DMT but decided to refrain because of safety concerns about the powerful magnetic fields of MRIs and radioactivity in PET scanning. Even without this information, we have a good idea of where DMT is active in the brain. The limbic system is the region primarily affected by psychedelics such as LSD and ketamine, which makes it a likely candidate for DMT, and the limbic system is involved in OBEs.

DMT is a derivative of serotonin, and is thought to act upon the serotonin receptors. Since the limbic system's serotonin receptors are sites for the psychedelic effects of drugs, it makes sense that DMT would have a similar effect. This is still consistent with Jansen's theory about the limbic system's NMDA receptors in OBEs and NDEs, because indirectly, via the serotonin system, DMT could affect NMDA receptors and lead to ketamine-like effects.

The body's natural source of DMT is the pineal gland, which is a centrally located pea-like structure.[33] DMT is released regularly during dreaming and may contribute to the bizarre qualities of some dreams. The pineal gland also makes beta-carbolines, chemicals that prevent destruction of DMT by enzymes in the brain. Those enzymes are in the brain in order to limit DMT's effects, so blocking them makes DMT more potent. Beta-carbolines are frequently added to concoctions with DMT, such as ayahuasca, in order to enhance its psychedelic properties.

So not only do we have a natural psychedelic in our brains, but there are times when the pineal gland releases beta-carbolines along with the DMT to make it even more active. Dr. James Callaway discovered some situations in which this occurs.[34] He detected an increase in a beta-carboline called pinoline when people were either dying, having OBEs, or experiencing lucid dreams. He also found that pinoline increases during stress in response to adrenaline, which may be why stress can cause OBEs.

THE PINEAL GLAND

Until fifty years ago scientists had come to regard the pineal gland as a vestigial organ with no current use, like the appendix. This opinion arose in part because the pineal gland becomes calcified as we age. The depositing of calcium in the pineal gland begins in adolescence, and anywhere from 33 to 76 percent of the pineal gland is calcified by the time of death, but the reason for the calcification is unclear.

But some scientists believed that the pineal gland must have an important function. Their belief stemmed from two facts: (1) the pineal gland is bathed by cerebrospinal fluid (CSF) to a greater extent than other sections of the brain, enabling the pineal gland's secretions to circulate quickly to the brain's other parts, and (2) the pineal gland has a significant blood supply, perhaps more blood flow per cubic centimeter than any organ other than the kidneys. If it were not extremely important to the brain and body, it wouldn't have such rich resources for receiving and sending chemical signals.

The proof that the pineal gland played more than an inactive

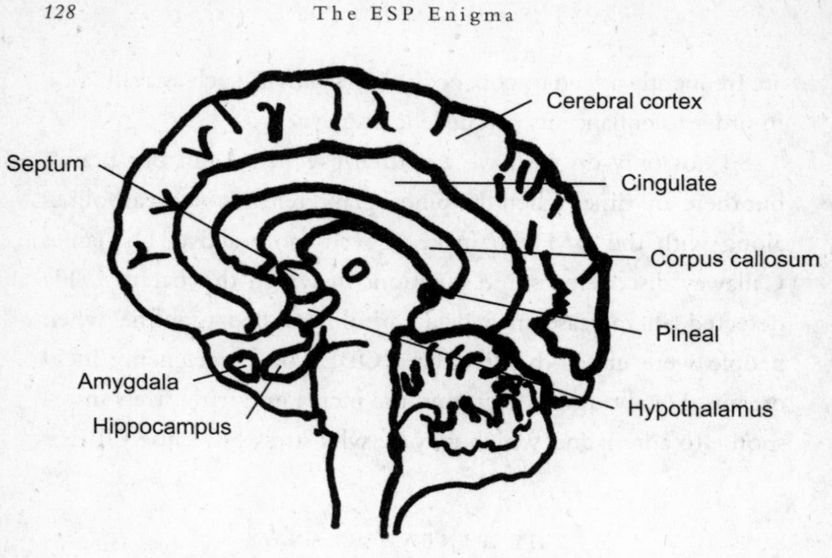

Figure 3. The pineal gland.

role occurred in the 1950s after Aaron Lerner discovered that it produced melatonin, the chemical messenger that induces sleep. Melatonin is made in the pineal gland from the amino acid tryptophan, and it is also synthetically manufactured to treat insomnia and jet lag. Light turns off the production of melatonin, and production resumes after dark. Despite being buried in the center of the brain, the pineal gland has maintained its sensitivity to light through its neural connections to the eyes' retinas.[35] And although melatonin is the primary sleep-related chemical secreted by the pineal gland, the gland also secretes arginine vasotocin (AVT), which induces slow-wave sleep. When antibodies to AVT were given to cats, their immune system reduced their brain's AVT and the cats spent more of their sleep time in REM (dreaming) sleep.

So the pineal gland is the master gland for chemically orchestrating the transition to sleep by its secretion of melatonin and

AVT. Atypical secretion of these substances may explain why some psychics have brainwaves consistent with dreaming while awake. And atypical secretion of DMT and pinoline by the pineal gland might cause OBEs and other experiences associated with psychic abilities.

Another way of assessing the pineal gland's involvement in psychic phenomena is to look at whether it is affected by some of the known influences on psychic abilities. One consistent observation is that psychic phenomena (telepathy, precognition, and clairvoyance) are strongest when the earth's electromagnetic field is at its lowest intensity. Indeed, the pineal gland is sensitive to the earth's magnetic fields. Mammals exposed to ELF (extremely low-frequency) electromagnetic fields develop an altered rhythm for melatonin synthesis. And the earth's geomagnetic influences on the pineal gland are one explanation for how birds navigate.

Some researchers have proposed that ELF waves are involved in the transmission of psychic phenomena. As mentioned, ELF waves affect the pineal gland. More importantly, ELF waves can elude the shielding of metal Faraday cages, because the cages eliminate only higher-frequency electromagnetic radiation. Also, the strength of ELF waves doesn't diminish as rapidly over distance as higher-frequency electromagnetic signals do, so they are a better candidate for transmission. But one major problem with the ELF hypothesis is that ELF waves are restricted to a very low-resolution signal. The data on psychic abilities suggest that we need to account for more detailed information than ELF waves could transmit. So psychic phenomena don't appear to be directly caused by electromagnetic transmission, even at low frequencies.

Evidence does support the idea that the pineal gland plays a role in creating states of consciousness that are conducive to receiving psychic information, such as dreaming and OBEs. These states enable us to experience information from the unconscious, which has a more expansive perspective of reality because it is not subject to the brain's filters. The information doesn't need to be transmitted to the unconscious. It already is there.

CAN CONSCIOUSNESS EXIST OUTSIDE OF THE BODY?

There is no evidence that consciousness is strictly confined to the limits of our bodies. In fact, there are reasons to believe the opposite. The nature of consciousness is still hotly debated, but no one thinks consciousness consists of matter. That means consciousness is either a form of energy or a type of field. Energy is defined as the capacity to do work, which doesn't describe consciousness as well as the definition of a field does. A field is what organizes or directs the energy or matter that is within its territory.[36]

Unlike matter, fields cannot be contained solely in the body. Magnetism and gravity both illustrate that fields exist in and around objects, not just within them. For example, the field around a magnet organizes iron filings placed around it. And the earth affects objects within its gravitational field, which extends far enough to keep the moon in its orbit.

If consciousness is a form of energy rather than a field, the skull is not an effective shield for preventing its leakage. The electromagnetic energy that is measured on our skull by an EEG does not stop there. It expands beyond our brain and weakens with distance. Regardless of whether consciousness is an electromagnetic field or another field, it probably extends beyond the physical body. However, it does not extend far enough for that alone to explain most psychic experiences.

The data from NDEs and OBEs take the discussion to another level by raising the possibility that consciousness can exist independent of the body and travel to other locations. There is no scientific correlate for an astral body that can travel from one's body to far-flung locations and back. But there is an alternative explanation for OBEs and NDEs—when people experience OBEs, or have psychic abilities where they navigate other locations in space and time, they could be accessing parts of their consciousness field that are normally out of conscious awareness. This would require our consciousness field to be coupled with the outside world, which the evidence suggests is true.

NDEs in which the person actually died and was revived are a strong argument for consciousness existing independent from a brain. The critical question is whether the visions and sounds of the NDE could have been accessed psychically immediately after the person was revived, instead of being due to consciousness continuing during the time of brain death. People often have newly acquired psychic abilities after an NDE that might enable them to fill in the gap for what had just happened.[37]

The possibility that people's brains would compensate for the gaps in consciousness during NDEs is not without precedent. Our brains do not like lacunae in information and try to fill them in. People with hippocampal damage confabulate rather than experience a gap in their consciousness. We all have blind spots that result from the locations where the optic nerves exit the back of the eyes. Rather than see black areas in our visual field, our brain uses surrounding visual information to fill in the holes, and the right temporal lobe plays a part in this process.[38] These gaps are filled in automatically and unconsciously, so we don't even know that they exist unless something unexpected moves out of our blind spot into our field of vision. Similarly, the survivor of an NDE could be unaware of the gap and psychically access the information, thereby creating the illusion of a continuous flow of consciousness.

There is a large body of research on the survival of consciousness after death that includes research on mediumship (contacting the dead) and reincarnation. The research is intriguing and compelling, and suggests that it is possible that consciousness can survive death. The problem is that all of the data can also be explained by psychic abilities. The medium could be psychically accessing information rather than contacting the dead. Similarly, the person who recalls a past life that is verified may be psychically accessing this information. Until data are collected that can get around this difficulty in distinguishing between the two explanations for NDEs, science must apply Occam's razor. But which explanation you personally prefer is up to you.

Chapter 8

EVOLUTION AND EXTRAORDINARY HUMAN ABILITIES

Let us learn to dream, gentlemen; then we shall perhaps find the truth.

—FRIEDRICH KEKULÉ

THROUGHOUT HISTORY, PSYCHIC DREAMS have been a strong and recurrent theme. The ancient Egyptians built temples for inducing prophetic dreams. Australian Aborigines believe that knowledge is first acquired during dream time. Edgar Cayce was called the "sleeping prophet" because of his trance-like state, in which he visited "all of time and space." Most spontaneous telepathic and precognitive experiences occur during dreams. And images transmitted telepathically are less fragmentary when they occur in dreams than when they occur in the awake state.[1] It would make sense for the brain circuitry involved with OBEs and psychic phenomena to overlap with that of dreaming.

But how about the people who have psychic experiences when they are awake and not having an OBE? A clue to their brain activity comes from medical syndromes with a high incidence of psychic abilities, such as synesthesia and autism. Studies of people with these syndromes reveal that their brain activity patterns while awake show similarities to those of other people while dreaming. This finding strengthens the link between psychic abilities and the brain's circuitry for dreaming. This link may result from common evolutionary roots. In other words, psychic abilities appear to be an evolutionary development that became partially buried but resurfaces in modern man when the evolutionarily newer areas of the brain are less active.

THE EVOLUTION AND PURPOSES OF DREAMING

Dreaming serves multiple purposes, including laying down long-term memory and processing emotional experiences. When one looks at the context under which dreaming evolved, psychic communication might have been another primary purpose. In the evolutionary tree, dreaming or REM sleep did not appear before mammals and birds. Reptiles do not have REM sleep, which means they do not dream. Birds evolved from reptiles independently from mammals, which means that they probably developed REM sleep on their own. Birds, such as pet parrots, have also been reported to exhibit telepathic behavior, which fits with the link between telepathy and the brain circuitry for dreams.[2]

The limbic system, which is important for REM sleep, is sometimes called the "mammalian brain." A cruder version exists in reptiles and birds, but it is most highly evolved in mammals, where it includes more structures and some areas of cortex. Dreaming (REM) sleep is present in all mammals except the egg-laying ones, and the only surviving egg-layers are the platypus and spiny anteater. This means that REM sleep probably emerged sometime after the split from the common ancestor for marsupial, placental, and egg-laying mammals but before the split from an ancestor common to placental and marsupial mammals. Methods such as radiocarbon dating of animal remains and comparisons of the genetic material of existing animals have allowed scientists to estimate when this evolutionary split in the ancestral tree occurred. It is estimated that egg-layers split off from other mammals approximately 150 to 210 million years ago and that marsupials split off from placental mammals 130 million years ago. This would place the evolution of REM sleep in mammals somewhere between 130 and 210 million years ago.

The limbic system performs several vital functions for mammals and birds that the reptiles preceding them didn't need. For example, mammalian and avian babies became more dependent upon their mothers for survival. These babies were also fewer in number, so their survival became more important. Survival of offspring was aided by the limbic system's promotion of social bonding and nurturing behaviors. And whereas reptiles and mammals will sometimes eat their own young, mammals differ from reptiles in that mammals will often risk their own lives to protect their young.

The fact that mammalian and avian offspring needed more protection for survival could also have been a driving force for the natural selection of psychic abilities. Jon Tolaas developed a theory that helpless newborns spend much of their time dreaming because that state lets them psychically detect threats and silently communicate them by telepathy to parent(s), even when they are outside ordinary communication range.[3] Mammals and birds spend a great deal of time sleeping as a way to conserve energy because, unlike reptiles, they are warm-blooded and require more energy pound for pound than cold-blooded animals do. Sleep became essential, but sleep is also dangerous. Parents can't protect a vulnerable newborn as well when they are all asleep. Also, if the parents have to wander off to obtain food, this often means leaving the newborn alone and at risk of being snatched by a predator. Offspring with dream telepathy would have an increased survival rate and more of their genes would get passed down to future generations.

Consistent with Tolaas' theory, newborns spend much of their time asleep, and a disproportionate amount is REM sleep. For example, an adult human spends a total of one and a half hours per twenty-four-hour cycle in REM sleep, whereas a newborn spends nine hours. This trend is greatest in those species that are most helpless at birth (humans, cats, rabbits, and dogs). Tolaas's theory ties in with the research observation that telepathic dreams occur mainly during times of crisis and between loved ones. In humans the psychic bond can be so strong that the crisis can be relatively minor. There are many anecdotes of mothers who awoke in the middle of the night without knowing why. They then checked on their children literally seconds before the children awoke crying from being sick or having a nightmare.

Other psychic abilities may also have coevolved with dreaming. According to Mark Mahowald, director of the Minnesota Regional Sleep Disorders Center, "almost the entire state of being before we're born is REM (dreaming) sleep."[4] Obviously, when a baby is in the womb, dreaming wouldn't be necessary to protect the baby from predators, but it could serve another purpose. Hypnotized people have accurately described traumatic events that occurred externally when they were in the womb. And some people have these kinds of memories surface after an NDE left them with enhanced psychic abilities.[5] It may be that these individuals psychically accessed the past. But these accounts also raise the possibility that a dreaming baby can access external information while within the womb.

This theory is supported by the fact that human newborns

can immediately distinguish their mother's voice from those of other women, even though the voice as heard from inside the womb would be very muffled compared to outside the womb. The newborn's ability to orient to the world outside might be aided by precognitive or clairvoyant dreams before birth. Dreaming before birth would help prepare the baby psychically for the shock of what lies ahead. The world outside the womb is full of sights, sounds, smells, and sensations that would be frighteningly alien to someone coming from a dark, fluid-filled sac where one's basic needs are automatically met. Yet babies adapt remarkably well to the world after birth. And an ability to preview the external world would be more important for mammals than reptiles because the limbic system renders mammals more emotional and vulnerable to being traumatized.

Another function of dreaming was proposed by Michel Jouvet of the University of Lyon.[6] He proposed that dreaming permits the testing and practicing of genetically programmed behaviors in sleep. Doing things in dreams is just like actually doing them as far as the body is concerned, so dreams can be extremely helpful in developing skills. Dreams are the perfect way for inexperienced baby mammals to learn safely, since they are paralyzed while dreaming.

This benefit of learning during dreams continues into adulthood. Jack Nicklaus claimed that he discovered a modified way to grip his golf club in a dream that improved his game by ten strokes overnight. Psychoanalysts regard many dreams as a substitute for actually fulfilling one's wishes, but they may be a means toward making one's dreams or wishes come true by helping us develop the skills we desire or need.

One would think that we'd remember our dreams more easily since they are instrumental to learning. Stephen LaBerge suggests that this forgetfulness may relate to an inability of animals to distinguish between dreams and reality. If a cat dreams that the vicious dog next door was killed by a car, it might venture over there the next day only to find a terrible surprise. So it is best that the cat not remember the dream.

The ability to distinguish between a dream and reality is acquired in humans between the ages of four and six. The distinction can become blurred again for various reasons, such as mental illness or a long stretch of sleep cycle disturbance. People report a higher incidence of psychic phenomena in those situations where dreaming and reality become more difficult to distinguish, but the phenomena are usually not taken seriously by psychiatrists. Psychiatrists are taught that claims of being telepathic or psychic are signs of mental illness, but this is often not the case. Also, even people who struggle to maintain a grip on reality may still have verifiable psychic abilities.

EVOLUTION OF THE HUMAN BRAIN

In order to understand how psychic abilities became buried and why they occasionally resurface, it is necessary to know about the evolution of the human brain. After the limbic system developed, the next phase in mammalian evolution was the increase in size of the outermost layer of the brain, the cortex. The highest level of processing sensory information occurs in the auditory cortex, visual cortex, olfactory cortex, and somatosensory cortex (which process touch, pain, vibration, hot and cold, and the position of body parts). We also have association cortices, which integrate information from the sensory cortices. The association cortices have been called "silent" because minor damage has not had obvious consequences, but these silent areas are thought to be where complex thinking occurs. These areas allow us to form abstract concepts about the world.

Human brains developed such an expanded cortex that it has a corrugated appearance, whereas the rat's cortex is very smooth. If one were to flatten out the corrugations, one would see that the human cortex is vastly larger than the cortex of most other animals, and bigger than one would expect by simply comparing our skull size to that of other animals. The brain's expansion occurred mainly between the appearance of *Homo erectus* 1.9 million years ago and archaic *Homo sapiens* 500,000 years ago. The brain doubled in size during that time, resulting in a brain only 15 percent smaller than modern man's. During the subsequent 425,000 years, the brain grew the remaining 15 percent, which led to us, *Homo sapiens sapiens*. The association cortices that expanded the most are in the temporal, parietal, and frontal lobes. They are often called "neocortices" because they are evolutionarily the newest.

Our frontal lobes grew to occupy close to one-half of our brain's total cognitive capacity, making us better able to anticipate the future, keep our impulses in check, set goals, and avoid dangers. The flip side was that our frontal lobes became a major source of neurotic behaviors and thoughts. They increased our capacity to acknowledge our own mortality. And they enabled us to worry: *Did I turn out the lights? Did I lock the door? Did I say the right thing? Will I get cancer? Will I find a job? Will I ever get married?*

Thinking takes up most of our conscious awareness, even though much of our thoughts are repetitive, mundane, and not very productive. Our conscious thoughts are primarily in the form of language, which usually involves the left brain.[7] These thoughts are responsible for creating the "monkey mind," which is the Buddhist term for the constant background thoughts or futile chatter that we silently engage in. Often meditation turns down this activity and the mind becomes quiet.

The development of the left brain is more recent than that of the right brain, and the left brain inhibits activity in the limbic system and right brain. The advantage of this dynamic is that the logical left brain can damp the emotional right brain and limbic system, enabling one to think before reacting. The reward or pleasure center is in the limbic system, so the left brain keeps some of our hedonistic impulses in check. The left brain also enables one to have a poker face when it is to one's advantage, and to rationalize away fears that might prevent one from doing something that has posed dangers before. But as a result of this inhibition, less information enters conscious awareness from our entire field of consciousness. This inhibition may keep us better-behaved and from being overwhelmed, but it also keeps us from being as tuned in to others or as aware of the world around us as we might otherwise be.

The right and left brains are asymmetrical in their distribution of the neurotransmitters, or chemical messengers. The right brain uses more norepinephrine, and the left brain uses more acetylcholine and dopamine. Norepinephrine is involved with arousal responses and fears, which may relate to nightmares and anxiety dreams, but also to crisis telepathy. Acetylcholine is necessary for cognitive functioning. Drugs that block it cause forgetfulness, and most treatments for Alzheimer's disease enhance it. Nicotine stimulates acetylcholine receptors and also enhances left brain functioning. Dopamine is a messenger for the pleasure center in the limbic system and a primary neurotransmitter for focus or concentration in the frontal lobes. Drugs such as Ritalin that increase dopamine can increase left brain functioning, but they also can be abused because of their effect on the limbic system's reward center. In addition to the use of drugs, genetic variations among individuals impact these chemical messenger systems and affect the dynamics between the left and right brain. These variables make it easier for some people to have more access while awake to certain areas of their consciousness field. But the dynamics between the left and right brain shift for all of us when we are dreaming, and widen our access even more.

DREAMING AND TURNING OFF THE ANALYTICAL MIND

Since psychic information is so strongly associated with dreaming, brain imaging of the dreaming state provides more clues to what brain activity is associated with psychic abilities. PET scans of people during the sleep cycle show the opposite activity pattern from when we are awake. The associative cortices are relatively quiet during REM sleep and our limbic system and sensory cortices are more active.[8] The left/right dominance is also shifted. Our sensory cortices provide the sounds, sights, smells, and so on of our dreams. The limbic system determines what is emotionally important, so our dreams reflect what we desire, worry about, or are struggling to understand.

Imaging studies during dreaming show that the activated sections of the brain are those that perceive or feel information rather than analyze it. Psychic information comes as a perception, rather than through logical analysis. Similarly, dreams tell us something by showing it to us, even though it may be in symbolic form. We usually don't question what happens during dreams because our analytical mind no longer dominates. So our dreams can contain illogical content, such as someone with three eyes or a flying elephant. This abandonment of rules about what is possible is why dreaming can be helpful in solving problems, because it doesn't have the biases that can prevent seeing the truth.

Dreams have allegedly led to discoveries that scientists were previously unable to find answers to through logical analysis. One was the discovery by Friedrich Kekulé of benzene's molecular ring structure. The prevailing theory at the time was that benzene had a linear structure, until his dream:

> I turned my chair toward the fire place and sank into a doze. Again the atoms were flitting before my eyes. . . . Long rows frequently rose together, all in movement, winding and turning like serpents. . . . One of the serpents seized its own tail and the form whirled mockingly before my eyes. I came awake like a flash of lightning. . . . I spent the remainder of the night working out the consequences of the hypothesis.[9]

Kekulé's dream reminds me of the experiences of people on DMT who see serpents give them information about medicinal properties of plants, or who see structures such as double helices, which is the structure of DNA, our genetic material. This similarity may be because some DMT is released from the pineal gland during dreams.

Dreams such as Kekulé's incorporate colorful and fanciful imagery. They are not meant to be taken literally, but can still lead to a symbolically masked discovery or truth. But sometimes more direct dream imagery is the only way to display the truth. That was the case for the discovery of the periodic table, the chart that organizes all of the chemical elements. Before 1869 around thirty elements were still undiscovered, and the absence of a clear pattern made it difficult to classify the known ones.

Dmitri Mendeleev, a Russian chemist, wrestled with the problem until one night: "I saw in a dream a table where all the elements fell into places as required. Awakening I immediately wrote it down on a piece of paper. Only in one place did a correction later seem necessary."[10] The resulting table was so accurate that the properties of unknown elements were correctly predicted. Mendeleev's dream is reminiscent of the autistic savants who just see the answer to a mathematic equation rather than work through the calculations. Not surprisingly, autistic savants have brain activity that shares characteristics with dreaming.

EXTRAORDINARY TALENTS

Autistic savants are a prime example of people who obtain information through heightened perception rather than the usual analytical channels. Savants demonstrate abilities that would be remarkable for someone with a genius IQ, but they are even more remarkable because savants often have low IQs and lack the education and cognitive abilities normally associated with their talents.

A pair of twins described in Oliver Sacks's book *The Man Who Mistook His Wife for a Hat* amused themselves by calling out six-digit prime numbers that just appeared in their minds. There is no algorithm for determining primes, so this would be inexplicable even if they had degrees in higher mathematics. They also had calendar-calculating skills that spanned more than eight thousand years. In other words, they could immediately tell you what day of the week September 4, 3012, would fall on. There are algorithms for calendar calculation, but many savants, despite being incapable of simple addition, can give the answer faster than a mathematician using an algorithm. Another example is the ability of some savants to almost instantaneously give an accurate count of hundreds of matches thrown on the floor.

Daniel Tammet is an autistic savant who can perform mathematical calculations at incredible speeds.[11] He became obsessed with counting shortly after experiencing an epileptic seizure at the age of three. He can determine cube roots of numbers as fast as a calculator, recall the mathematical constant pi to 22,514 decimal places,[12] and he can speak seven languages.[12] But because of his autism, he can't tell left from right (a left brain function) and has difficulty with simple tasks such as going to the grocery store.

Kim Peek, the man upon whom the movie *Rain Man* was based, can read two books simultaneously, one with each eye, and recite in detail the twelve thousand books he has read, both backward and forward.[13] Stephen Wiltshire is an artistic savant who drew a highly accurate map of the London skyline from memory after a single helicopter trip. Leslie Lemke is a blind savant who played Tchaikovsky's Piano Concerto No. 1 after he heard it the first time. Like most musical savants, he never had a piano lesson.

In addition to having the typical savant skills in mathematics, art, music, and memory, some savants have psychic abilities.[14] George was able to tell in advance when his parents would arrive to pick him up at school, even though they provided no warning and the more typical scenario was for him to take the bus. Michelle had several dozen episodes of clairvoyance, one of which was when she told her father that he had replaced a watch crystal that had fallen out in the bathroom. She said this shortly after it happened and without any way she could have known. Ellen was a blind musical savant who could predict her Christmas presents one week in advance without any clues. Another savant dreamed that her father died from a heart attack the day before he died that way.

No one has understood how the savants perform their feats, but the "savant syndrome" is over a hundred times more prevalent in autism than in other forms of mental retardation or mental illness, and almost 10 percent of autistic individuals have some savant skill(s). What is it about autism, which otherwise severely impairs functioning, that can lead to such seemingly superhuman abilities? One clue comes from the triad of symptoms for diagnosing autism (delays or aberrations of communication, poor social skills, and odd or repetitive behaviors). These suggest dysfunction of the left hemisphere (language) and the limbic system (social bonds and the ability to adapt to situations).[15] Autism may give rise to the savant syndrome because it has the same shifts in dominance described for dreaming. Bernard Rimland of the Autism Research Institute in San Diego maintains a database of more than 34,000 people with autism. He noted that the savant skills are usually enhanced right brain functions and savants' deficits are in left brain abilities.

Most autistic individuals do not have the language skills to describe their inner experience, but some do. Temple Grandin, the high-functioning autistic professor of animal science who coauthored *Animals in Translation*, says that she "thinks in pictures" and doesn't "abstractify." This suggests a shift to right brain functioning. It gives her an advantage in her work because people who think abstractly see what they expect to see, rather than what actually is. Grandin says she sees what animals see and can design better equipment for handling them as a result.

The ability to form abstract concepts involves the neocortices, particularly in the frontal lobes of the brain and on the left, whereas the right brain and visual cortex are involved in "thinking with pictures." Grandin says that the world she sees is full of details. However, this has its downside. Abstract concepts allow one to generalize and reduce the amount of detailed information that needs processing. Too much information can become overwhelming and make it difficult to function.

Functional MRI studies of autistic individuals and IQ-matched controls are consistent with the proposed shifts in dominance. When both groups were given identical memory and attention tasks they performed at equal levels, but used different sections of their brains.[16] The controls activated several areas of their left and right neocortices in an integrated fashion, whereas the autistic subjects preferentially activated a small portion of their right neocortex or both sides of their visual cortex. This fits with the evidence of damage to the left brain in autism, and also with how Grandin describes her thought process as primarily visual.

In dreaming and autism there is a shift in dominance, so the limbic system and sensory cortices are more dominant than the neocortices, and a shift from the usual left brain dominance to right brain dominance. The two shifts are not independent of each other, because increased activity in the limbic system will increase the activity of the intuitive, creative right brain and inhibit the analytical, linear, and logic-based left brain.

The shift from left to right brain functioning was an expected finding, but the degree of disconnection in the brains of autistic individuals was not. The current model of brain functioning had led to the expectation that individuals with savant skills would have greater or more complex connectivity within their brains' circuitry. However, rather than having more connectivity, these studies and others show that they have less. In fact, Kim Peek has no corpus callosum, which is the band of fibers that connects the left and right brains. This may be why he is able to read two books simultaneously. Also, because the left brain inhibits the right brain through the corpus callosum, this finding suggests that savant skills might be assisted when the left brain can't interfere with the right brain.

Further evidence that damage to the neocortex assists savant abilities comes from another neuropsychiatric disorder: frontotemporal dementia (FTD), in which the temporal and frontal lobes deteriorate. Frontotemporal dementia is a disorder that, like Alzheimer's disease, causes the loss of cognitive abilities. University of California at San Francisco neurologist Bruce Miller noticed that some people with FTD suddenly develop aptitudes for art and music when they lose their language abilities. One of his patients had been a linguist but suddenly felt that his mind "had been taken over" by notes and intervals. He began composing music after the onset of his dementia, despite having had little musical training. Like autistic savants, Miller's patients became obsessed with their new pursuits and more attentive to details, patterns, and sounds. Brain scans showed that the left temporal lobe was damaged and less inhibitory of the right brain. Miller concluded that the dementia was not creating the talents but uncovering them.

More evidence for brain changes that lead to savant skills comes from the work of Allan Snyder, a neuroscientist at Australian National University. He used transcranial magnetic stimulation to disrupt the activity in the left frontal lobe of twelve normal volunteers. Depending upon how it is applied, magnetic stimulation can increase or decrease activity. In this case it decreased activity. For an hour or so after the magnetic stimulation, ten of the volunteers showed an improved ability to accurately guess the number of dots flashed on a computer screen. Although not as impressive as the savants who can count the number of candies in a jar, the results are consistent with the research on savants.

THE UNCONSCIOUS AND THE DEEP BLUE C: SYNESTHESIA

Synesthesia (from the Greek *syn-*, "together," and *aesthesis*, "perception") is a condition whereby sensory information is combined by the brain in unique ways.[17] A synesthete might always see the color blue when she hears the musical note C, red when she hears D, and so on. Such combinations can occur between any of the senses in synesthesia. They can also occur between an abstract symbol, such as a letter or number, and sensory input, such as color.

Most of us experience various pieces of sensory information as independent from each other, because our five senses enter our brain through different sensory organs and travel along sensory pathways that are completely separate from one another. Our brains then combine information from these sensory systems into a coherent package that we experience as the external world. However, the different sensory pathways are very close to one another at the angular gyrus.

The angular gyrus is the area that creates partial OBEs when stimulated and showed signs of damage in people who had experienced NDEs. The neurologist V. S. Ramachandran proposed that certain types of synesthesia result from "cross-talk" between the different pathways that run through the angular gyrus.[18] Ramachandran also described a case of a color-blind synesthete who saw colors associated with numbers. Colors he normally couldn't see were evoked by certain numbers. The colors were so alien to the synesthete that he referred to them as "Martian colors." This is reminiscent of individuals who are blind from birth but can see during their NDEs. People who become blind after birth can see during regular dreaming, but those with blindness at birth can access visual information only through an OBE that allows exploration of a consciousness field that is normally walled off.

The neurologist Richard Cytowic found that 17 percent of synesthetes have a high frequency of "unusual experiences" such as clairvoyance and premonitions. Several of the psychics I have encountered are synesthetes with two or more linked sensory systems. Many of them see auras (colorful fields of light around the body). Seeing auras, reported by 3 percent of synesthetes, is thought to be a form of synesthesia in which colors are paired with personality types.

P. M. H. Atwater, the near-death-experience researcher, had psychic abilities and several senses interconnected before her NDEs. She sees music, hears numbers, and smells colors. Also, some of the autistic savants, such as Daniel Tammet, have synesthesia. He attributes his ability to remember the digits of pi to the fact that they are all associated with colors. In order to remember pi he just visualizes a stream of colorful images associated with the numbers.

Synesthetes often have an incredible fund of knowledge. Like children who survive NDEs, synesthetes generally have superior scores on IQ tests. Their memories can be spectacular. *The Mind of a Mnemonist* is about a synesthete (given the pseudonym S) who earned a living by displaying his ability to remember series of numbers and words.[19] His memories appeared to be limitless and came effortlessly. In fact, S described his experience of recall as "the thing remembers itself." His memory was so powerful that it often rendered him dysfunctional in other ways. The amount of information was overwhelming. He could remember in detail everything said or done on a specific day chosen randomly from his life. Remote influence or psychokinesis was another skill he experienced. He said, "If I want something to happen, I simply picture it in my mind. I don't have to exert any effort to accomplish it. It just happens."

Like people with autism, synesthetes often have clinical signs of left hemisphere damage and right hemisphere dominance. These signs include subtle mathematical deficiencies, confusion of their left and right, and being left-handed or ambidextrous. (The left hand is controlled by the right brain and vice versa. The majority of right-handed people are so because of genetics, but a very tiny number become right-handed from right brain damage. Most left-handed people inherit a gene that leaves their handedness up to chance or another unknown factor, so it can run in families but skip generations. The percentage of left-handed people whose handedness results from brain damage is much greater than that of right-handed people, as evidenced by a greater presence of neurological symptoms of brain damage in left-handed people.)

The first imaging study on synesthesia was done by Cytowic two decades ago. The subject had an unusual form of synesthesia: he felt spheres, cones, columns, and other shapes press against his skin whenever he smelled different fragrances. The imaging study used an old technology wherein the subject inhaled radioactive xenon gas to show which brain areas were the most active. The results were shocking to Cytowic. The outer layer of the subject's brain (neocortices) had almost completely shut down. He had low blood flow in both the left and right brains to begin with, but his left brain's activity dropped another 18 percent when he smelled the fragrance. His brain activity was so low that a neurologist reading the study would expect the subject to be blind or paralyzed, but he had no neurological or cognitive deficits. In fact, he had an IQ of 130.[20]

Other studies of blood flow in the brains of synesthetes have shown that the limbic system, particularly the hippocampus, is more active. Also, limbic seizures and electrical stimulation of the limbic system can cause synesthesia experiences in nonsynesthetes. The limbic system is where LSD primarily works, and Grossenbacher noted that "if you give people enough LSD or mescaline, they will often experience synesthesia."[21] The limbic system is overactive in depression, and McKane and Hughes described two cases of women who developed synesthesia at the onset of their depressions. The synesthesia disappeared when their depressions responded to antidepressant therapy.[22]

COULD SAVANT SKILLS BE RELATED TO PSYCHIC ABILITIES?

People such as S and Daniel Tammet have such incredible recall that it challenges what neuroscientists propose as the mechanisms of memory. According to neuroscientists, memory is a function of enhanced synaptic connections in the brain's neural network. But people with autism have less brain connectivity, and some people with synesthesia have reduced cortical activity. Since both synesthetes and people with autism have a higher propensity for psychic abilities, as well as extraordinary memories, dramatic memory recall might be a psychic phenomenon. In other words, what if S accessed information from his personal past by a mechanism similar to that used by psychics to see into the future?

This capacity for memory may apply to all of us. Some memory, such as riding a bicycle, is probably encoded in our body as motor memory and requires no conscious attempt at recall. That is a different type of memory than what is used to remember facts we learn in school and events that happened in our lives. People can remember things under hypnosis that they had long forgotten, so we have a greater capacity for memory than we realize. In our normal waking consciousness, the dynamics of our brains give us a narrow access to our consciousness field. But we can make that change. One of the most remarkable qualities of our brains is their plasticity, or ability to form new connections and patterns of activity. By engaging in practices that shift our usual patterns of brain activity, we may find our inner savant.

Chapter 9

THE COMPARTMENTALIZATION OF CONSCIOUSNESS

The main theme to emerge . . . is that there appear to be two modes of thinking, verbal and nonverbal, represented rather separately in left and right hemispheres respectively and that our educational system, as well as science in general, tends to neglect the nonverbal form of intellect.

—Roger Sperry (1973)

OUR BRAINS COMPARTMENTALIZE INFORMATION into that which is conscious and that which isn't. By shifting our brains' dominant activity from the left brain to the right brain and from the cortex to its underlying subcortical brain, we can access information normally unavailable to our consciousness, such as psychic information.[1] Why these shifts have this effect becomes more understandable once you know the functions of these brain regions. Knowing these functions also provides clues to how you can make these shifts in your brain activity.

LEFT VERSUS RIGHT

In Western culture, left brain functions are valued much more than right brain functions. In fact, doctors initially concluded incorrectly that the right brain didn't do much. This opinion about the right brain arose in the early 1800s when Arthur Ladbroke Wigan did an autopsy on a man with whom he had been speaking just before his death. Although the man had been speaking rationally, he was missing his entire right brain. Wigan was initially shocked, but ultimately concluded that the right brain must be insignificant. Scientists didn't know at that time that the brain can compensate quite a bit when brain injuries occur early in development. In order for this man to have appeared so normal before death, the abnormality must have occurred during development, and some right brain functions were performed by his left brain.

It took scientists longer to understand the functions of the right brain than the left brain because language is primarily located in the left brain, which means that left brain damage is more apparent than right brain damage.[2] The right brain actually has several functions. Some are unconscious, and others relate to internal experiences that are not readily observable. As the table below illustrates, they are usually complementary to those of the left brain.[3]

Left Brain	Right Brain
Objective	Subjective
Logic-based	Feeling-based
Detail-oriented, sees parts	Big-picture-oriented, sees the whole

Facts rule	Imagination rules
	Source of dreams and day-dreams
Literal meaning	Metaphor, symbolism
Language, syntax, and semantics	Phonology, intonation, context, meaning, facial expressions
Future and past	Present
Math and sciences	Philosophy and religion
"Reality"-based	Fantasy-based
	Can be associated with drug-induced hallucinations
Strategies	Possibilities
Practical	Impetuous
Safety oriented	Risk taking
Vertical columns of neuronal connections	Horizontal axial connections
Serial or linear processing	Parallel processing
Dominant neurotransmitters are dopamine and acetylcholine (fine motor control and dexterity, speech)	Dominant neurotransmitter is norepinephrine (arousal to novel stimuli, visual-spatial perception)
Belief in being a separate individual	Feels a sense of unity with a higher power, unable to express why
Is purposeful, directed	Can enjoy just existing

Having a brain divided into left and right halves is not unique to humans; it is present in all mammals and many birds. The division has provided the ability to do two tasks at the same time. For example, many birds have their song function assigned to their left brain, which enables them to sing with their left brain while defending their territory with their right brain.

The separation and specialization of the two brains became necessary when their fundamentally opposite tasks required different, and incompatible, wiring from each other. The incompatibility is analogous to the inability to have programs in MS-DOS and Windows on the same computer hard drive. They would interfere with each other's ability to run.

Not surprisingly, there are differences between left and right brain circuitry. Some differences can be seen under the microscope. It is as though the left uses "in-line" circuits, like the string of Christmas lights that won't work if one bulb is dead, while the right uses parallel circuits, like the string of lights on which the rest of the bulbs work even when some don't. Parallel circuitry allows for much faster processing of information than in-line circuitry, just as a picture can relay more information in a short period than a verbal description can. The right brain's faster processing is essential for visual-spatial processing, gestalt reasoning, and determining the context or big picture. This is not to say that the right brain's functioning is superior or sufficient. The left brain's in-line circuitry is slower, but it enables information to be separated, analyzed logically, and expressed in a verbal fashion.

A band of 300 million nerve fibers called the corpus callosum enables the smooth transfer of information back and forth between the two brain halves.[4] The corpus callosum is missing in the autistic savant Kim Peek, who can simultaneously read a book with each eye. Most people have an intact corpus callosum, which makes this feat impossible. This is because of a phenomenon called binocular rivalry, which occurs when one eye looks at a different picture than the other. The brain can't see both pictures at once, so it has to choose between them. Sometimes one image is seen and the other is ignored throughout the testing, but often conscious awareness alternates between the two images. Brain imaging techniques are beginning to explore binocular rivalry to understand individual differences and why some visual stimuli result in a single image and others don't. This technology could be used to test people with highly developed psychic abilities and compare them with the general population to see if they show differences in binocular rivalry. For example, would people who are psychic be more likely to see both images, like Kim Peek?

THE SHIFT FROM A RIGHT BRAINED CULTURE TO A LEFT BRAINED ONE

Western culture has become increasingly left-brain-dominant, which correlates with its decrease in the intuitive or psychic way of obtaining information. Our language has been a contributing factor to this change. Some evidence for this theory comes from a study that showed that literate Greeks use the right ear and activate the left brain when listening to words. In contrast, illiterate Greeks use their left ear and activate their right brain more when listening to words. Just the act of learning to read influences which ear and side of the brain is used for listening.

The degree to which written language has influenced which side of the brain is emphasized depends upon the type of written language. Ancient written languages were pictorial and therefore used the right brain. However, as the number of words increased, there were more pictographs to memorize. For the sake of expediency, the Greeks developed our abstract alphabet between 1100 and

700 B.C. This shifted how our brains processed written language to the left brain, where phonetic language is processed. Languages that leave out vowels, such as Hebrew, require that the right brain still be very active along with the left in order to provide the proper context (right brain function) for understanding the text (left brain function).[5] For example, the right brain provides the context necessary to determine whether "s y sn?" means "See you soon?" or "So you sin?" Another cultural variation can be seen in Japanese, which has both a pictorial written language (kanji) and a phonetic script (kana). The right brain is used for kanji and the left for kana.

There is an additional reason why the three types of written language (pictographic, phonetic with vowels, and phonetic without vowels) activate our brains differently. Almost all pictographic languages are written vertically, while all phonetic ones are horizontal. Out of the several hundred languages with vowels, almost all are written left to right, whereas all those without vowels are written right to left.[6] Reading left to right favors the left brain, which contributes to modern Western languages' promotion of left brain dominance.

Left brain processing has also been emphasized by changes in how we measure time. For example, an hourglass represents time's passage by sand falling from its top half to its bottom one. When the sand hasn't entirely passed to the other half, figuring out the amount of time that passed requires comparing each half's volume of sand. This requires the right brain for interpretation. When cultures switched to analog clocks, the right brain was still necessary, especially if the hands pointed to positions and not numbers. Numbers are read, which is a left brain function because it is similar to reading letters. However, analog clocks still require right brain function to determine, with a glance, how much of an hour is left. Also, people with extensive right brain damage will ignore the left side of their world, including the left half of a clock. When clocks shifted to a digital readout of time, the time devices only required the left brain. They have become so common that many children in the West have no clue how to read an analog clock.

The Western world's educational system is geared toward learning by reading, which has further reinforced left brain dominance. The tendency to use the left brain over the right brain is so pervasive that the left brain is often used when the right brain should be. For example, adults who want to draw often need to be taught to use the right brain. They have to undo the tendency to use the left brain, which distorts the image because it is not the side of the brain for visual-spatial processing.

The biggest price we pay for this left brain dominance is in creativity, which decreases as we move through the years of schooling. Most American children test high in creativity prior to school. The percentage drops to 10 percent by age seven and to 2 percent by the time we are adults. This is relevant to psychic phenomena because creativity has been associated with increased psychic abilities.

Overdependence upon the left brain doesn't mean that we don't use the right brain; in fact, we continuously use the right brain in conjunction with the left brain. People with right brain damage illustrate the extent to which we use the right brain.[7] Right brain deficits differ depending upon where the damage occurred and include difficulty assessing the emotions of others, difficulty recalling emotional stories, an inability to place things into proper context, difficulty holding on to several possible meanings in conversation, a tendency to take things literally, and difficulty distinguishing between lies and jokes.[8]

There are ways to shift the dominance away from the left brain.

For example, Zen masters use koans, or nonsensical riddles, to push their students into using their right brain instead of their left. A well-known koan is "You can make the sound of two hands clapping. Now what is the sound of one hand?" The logical left brain may wrestle with the koan for a little while, but it will eventually resign and let the right brain grapple with it. Other methods for activating the right brain are staring at mandalas, dancing, drumming, and chanting. Some of these methods work because low sound frequencies preferentially activate the right brain and high sound frequencies preferentially activate the left.

Meditation synchronizes the activity of the two hemispheres such that the left brain is no longer dominant. EEGs of experienced meditators show more coherent, or lower-frequency, brain waves when they are meditating. This means that more brain cells within each hemisphere are in sync with each other. The activity of their cortex decreases during meditation and is less dominant over the underlying brain. So both the left-right and cortical-subcortical shifts from the brain's usual activity are present during meditation. And when experienced meditators were used for psychic research, such as remote viewing, their EEGs showed these same changes (greater coordination between the two halves of their brains and lower brain wave frequencies).[9]

HIGHER AND LOWER SENSORY PROCESSING

The division between the left and right brains, with the left brain dominant, is only one way in which information is segregated from conscious awareness. Another filtering mechanism occurs as sensory information goes through its multiple levels of processing before reaching conscious awareness. The cortex is the outermost section of the brain and is associated with high-level processing. The inner core of the brain is associated with low-level processing, which doesn't reach conscious awareness unless it is forwarded to the cortex.

We still can respond to something that doesn't reach conscious awareness. This is illustrated by pathological conditions such as blindsight, which results from damage to the visual cortex while the lower levels of visual processing are still intact. One patient with this condition couldn't see anything on his left side but was still able to distinguish crosses from circles and horizontal lines from vertical ones in his area of blindness. When asked how he did so, he said that he just guessed. This is how many subjects experience their answers during psychic research, even when they are scoring above chance. Psychic information, like perceptual information, first enters below the level of conscious awareness. Unless information reaches our conscious awareness, we don't know that we know it. But all of us can, and actually do, act on unconscious information.

"Flow" is a good example of how we can act outside of our awareness. Mihaly Csikszentmihalyi, former head of the department of psychology at the University of Chicago, is a leading researcher on flow, which he defines as being so absorbed in what you are doing that a euphoric sense of clarity and purpose takes over.[10] Flow is most likely when the task is neither too easy nor too hard. When in flow you act out of an unconscious awareness that is slightly ahead of your conscious awareness. The lower processing centers receive the information first, and during flow you take action before this information goes to the higher levels of processing. If you've played a lot of tennis, you may have experienced flow when you returned a serve before you consciously saw it.

There are times besides flow when we can act outside of conscious awareness. You may have stopped yourself from changing lanes without knowing why, only to have a car whiz out of your blind spot to pass you. Your lower level of visual processing saw the car move out of your blind spot before you consciously did. This ability to avert disaster without a conscious decision to do so is very adaptive.

Flow has been compared to meditation since both tap into an awareness that exceeds one's usual conscious awareness. People are so focused in both states that they lose a sense of self as separate from their surroundings. They become "one" with what they are perceiving or doing. Imaging studies of people during meditation and flow confirm that there is a shift in activity to subcortical centers.[11] The loss of a sense of self correlates with reduced activity in their parietal lobes, which are involved in sensing their body's relationship to the surrounding physical world. The increased singular focus on something in both states correlates with more activity in the prefrontal lobes, which are involved in attention.

One subcortical area for processing information is the thalamus, often referred to as a relay center for sensory information. Large cats with wide territories have specialized thalamic cells that are activated when they perceive something that moved into specific regions of their terrain. The activated brain cells are specific for precise geographic sites and are not dependent upon the cat's head position or viewpoint, which means the cat has a three-dimensional map of its territory within its consciousness field.[12] "It is almost as though the cat brain has on board a global positioning system (GPS)," noted James Austin, M.D., in his book *Zen and the Brain*.[13] Similar cells have been located in monkeys' hippocampi, the limbic system structures involved in memory.[14] That

Swann and Price could identify locations in remote viewing trials simply by addresses or geological coordinates suggests that humans may also have an innate but underutilized sort of "GPS."

The lower processing centers are called "lower" because they are deeper in the brain, evolved earlier, and receive sensory information earlier. However, they are not lower in any qualitative sense of the word. They handle a tremendous amount of information during ordinary perception. Whenever we look at something, our brains have to make trillions of calculations within milliseconds with visual information to create the image. Our brains determine the depth of objects by their shadows and the subtle differences between the images seen by our two eyes. We see thousands of different color hues even though our retinas only have three types of cones, each of which represents a primary color. All of those calculations are done without conscious awareness. And this is just a fraction of the information processed unconsciously. Whereas we only see an image as a result of trillions of unconscious calculations, autistic savants can also instantaneously attach a number to that image, such as the number of jelly beans in a jar.

THE UNCONSCIOUS

> *The eyes only see what the mind is prepared to comprehend.*
>
> —Henri Bergson

A simple definition of the unconscious is that it is everything in the mind that isn't conscious. The unconscious was once thought to be just a minor component of our mind, but it is now known to contain far more than our conscious mind. It also plays the largest role in determining our behavior. Our personal unconscious contains information from our past and present that isn't relevant to what we are currently doing, as well as things we don't want to face or admit to ourselves. Dreams provide us with a window onto our unconscious, and psychic dreams show that our unconscious can contain information beyond what is personal. How that is possible will be explored later, but first let's look more at perception and how the brain filters, processes, and controls what reaches conscious awareness.

Our unconscious mind is estimated to take in more than a billion pieces of information per second. This would quickly overload our conscious mind if it all reached consciousness, so less than 1 percent does. And how does the brain choose which 1 percent to let through? Much of the selection has to do with priorities about what information we need to function. We don't consciously determine most of these priorities. They are a product of the way our sensory organs and brain developed.

Since our most vital need is to respond to the demands of the moment, it is only in our dreams, OBEs, and altered states of consciousness that we totally lose our orientation to the here and now. That's why it's so hard to gain access to psychic information during normal waking consciousness. Even when our minds wander off to fantasize about some other place or time, we aren't completely removed from our orientation in space and time; we still operate in the present, albeit often on automatic pilot. So we can think about a vacation we took last week while we take a shower, wash the dishes, or do something else that requires only minimal attention. We can do these things with divided consciousness, which works well only when the tasks are routine. Fantasizing is completely different from the altered states of consciousness that allow access to psychic information, because it doesn't have the cortical-to-subcortical shift in brain activity needed for filtering mechanisms to be reduced.

THE BRAIN'S FILTERING MECHANISMS

The brain filters out psychic and other extraneous information from reaching consciousness by the following mechanisms.

Limitations of Our Sensory Systems

Our brain and sensory organs (ears, eyes, etc.) limit our perception of the physical world. For example, there are species with a greater range of smell, ability to see ultraviolet, ability to detect seismic forces before we do, and ability to hear frequencies outside our range. Some information in our world is only detected by man-made technology, such as the wireless devices that receive and send information signals through the air around us. These devices illustrate a previously unthinkable capacity for information storage in what we can only perceive as empty space.

Interacting with the World at the Macro Level

We see and touch what appears to be a very solid material world, but physics has found that atoms consist primarily of space. Even though objects obstruct our view and we know we can sit on chairs without falling to the ground, science tells us that these objects are predominantly space because they are made of atoms. But we couldn't interact with objects if our brains were wired to focus on the space within atoms rather than their minor material part. What we consciously perceive enables us to perform our day-to-day interactions, but our unconscious does not need to be limited to what is pragmatic. The unconscious can experience the "impossible" during OBEs, precognitive dreams, and other psychic phenomena.

Thresholds

Information has to pass thresholds to reach conscious awareness. An example is subliminal messages, such as the picture of popcorn that elicits craving for the buttery snack when flashed on a movie screen during a film. The image doesn't reach conscious awareness because it is visible for only a thousandth of a second and we filter out anything that lasts less than one-sixteenth of a second. But the picture increases sales of popcorn because the information is still perceived by the unconscious and is influential despite being out of awareness.

Latent Inhibition

The brain's filtering process includes what psychologists call "latent inhibition."[15] This term refers to the reduction of attention to stimuli that historically have been inconsequential. We automatically tune out the constant hum associated with our car running properly, but we'll react if the engine makes a loud clunk. We also feel our clothes most when we first put them on, but within minutes we feel them less. People differ in the degree to which they tune out background noise. Research has found that people who are open-minded, creative, or intelligent score lower on tests of latent inhibition, which means they tune out less information.

This lower latent inhibition is also correlated with increased psychic abilities.

Interest

Most of us are caught up in our own concerns and interests, particularly in these hectic times. By focusing on what we are interested in, we filter out information we don't care about and don't want as a distraction.

Expectations

A major factor in how we perceive the world is how we've learned to experience it, which leads to our habits and expectations. As a consequence, our conscious minds both miss and misperceive things much of the time. As Yogi Berra said, "If I hadn't believed it, I wouldn't have seen it." Berra's statement may apply to psychic information. We don't expect to be able to know the future or what someone else is thinking. This may be why researchers consistently find that their results are more likely to favor psychic abilities if the subjects believe such abilities exist, even if the subjects have no history of psychic experiences.

This effect of expectation on everyday perception is demonstrated by a phenomenon called "change blindness." There can be large changes in a scene that the viewer fails to detect when they occur unexpectedly.[16] Experiments have had subjects talk face-to-face with a stranger. The subjects will fail to notice that the stranger was surreptitiously replaced with a different actor if the switch occurred during a very brief diversion. Even if the actor looked very different, if he or she continued the conversation as though nothing changed, the subject didn't consciously perceive the difference. Since the subject's unconscious perceived the change, the subject might have a strange feeling without knowing why. People with autism are very sensitive to changes in their environment and do not filter out details because of expectations, so they don't experience change blindness or optical illusions.

"Abstractification"

Most of us distill details perceived at an unconscious level into an abstract concept. Our ability to communicate verbally relies upon this. People with autism see the world without forming abstract concepts, which makes it hard for them to communicate. Because of this inability to "abstractify," Temple Grandin has said that she has no unconscious—that her conscious awareness contains what is normally unconscious. The unconscious is out of awareness by definition, so one can't know if it doesn't exist, but it's fair to say that Grandin has access to more of what is normally unconscious than most of us do. And some savants have shown us that the unconscious can contain far more than we ever imagined. When one considers all of the brain's filtering mechanisms, it becomes apparent that we are consciously aware of only an extremely narrow window of reality.

Chapter 10

CONSCIOUSNESS AND THE WEB OF LIFE

> *All things are connected like the blood which unites one family. Whatever befalls the earth befalls the sons of the earth. Man did not weave the web of life, he is merely a strand of it. Whatever he does to the web, he does to himself.*
>
> —CHIEF SEATTLE

OUR CONSCIOUS MIND doesn't see how interconnected the world really is, but sometimes we experience interconnection through strange coincidences. You might turn on the radio and hear it play the same song you were listening to in your head beforehand, even though it's been years since you'd thought of that song. If it only happened once, you wouldn't think much of it, but what if it happened several times in the same week? Understanding how such coincidences can occur against statistical odds requires that we explore the physics that underlie our interlinked universe.

People and objects appear separate and distinct from us and each other in our visible universe. We travel through three spatial dimensions (up-down, north-south, and east-west) and a fourth dimension, time. Unlike the spatial dimensions, time appears to move in only one direction. It goes forward and not backward, even though physicists say there is no physical law for why this should be the case. Instead, physicists propose that our concept of time is an illusion. They also say that our universe is highly interconnected, or that there are hidden relationships between things that appear separate. Physicists also propose that there are more than the four usually perceived dimensions. Our view of reality differs considerably from that described by physicists, but this is primarily because we are limited by our brains and sensory systems.

Psychic phenomena are clues that the physicists' abstract version of reality isn't just a mathematically derived concept, but one that may be closer to the way things actually are. That's because psychic phenomena are more possible in the physicists' version of the universe than the one we consciously perceive. Other phenomena, such as synchronicities and mystical states, also lend support to modern physicists' theories about reality. I'll present them briefly before discussing the relevant physics.

SYNCHRONICITIES

Carl Jung coined the term *synchronicity*, which means a coincidence that appears meaningful because the odds are extraordinarily high against it being due solely to chance. The following is an example from a Swiss journal that Jung relayed in a letter to H. J. Barrett on March 26, 1957:

> A man [was] celebrating his birthday; his wife had given him a new pipe as a present. He took a walk, sitting down on a bench under a tree. Another elderly man came along and sat

> down beside him, smoking the same kind of pipe. Mr. A. drew Mr. B.'s attention to the fact that they both smoked the same pipe, whereupon Mr. B. told him that he was celebrating his birthday on the same date and had received the pipe from his wife. He introduced himself and it turned out that both had the same Christian name Fritz.[1]

Another synchronicity that caught Jung's attention involved the French writer Émile Deschamps. A man named Monsieur de Fontgibu had given Deschamps some plum pudding sometime in 1805. Ten years later Deschamps saw plum pudding on the menu of a restaurant in Paris. He attempted to order some, but the restaurant was out of plum pudding because the last dish had been served to Monsieur de Fontgibu. Then in 1832 Deschamps visited another restaurant with a friend, saw that plum pudding was on the menu, and told his friend that story. He said that if Monsieur de Fontgibu was there, "it would make the setting complete." At almost precisely that instant Monsieur de Fontgibu entered the room.

Both of Jung's stories have extraordinary coincidences without any apparent cause-and-effect relationship. The coincidences don't appear to have a hidden meaning, but they don't appear to be entirely random either. Many of us have experienced or heard about synchronicities because they are very common. An explanation for many synchronicities is that they are manifestations of the universe's underlying interconnectivity.

Alan Vaughan, the dream researcher who kept diaries to look for precognitive dreams, wrote a book called *Incredible Coincidences: The Baffling World of Synchronicity*. One story was of a woman in Berkeley, California, who was locked out of her house. She was wondering how to get in when her postman walked up and handed her a letter from her brother. There was a spare key inside. Another story was of a housewife who lost her ring in a potato field and found it forty years later in a potato she had just cut.

One of my patients flew to Hawaii just as her best friend was flying back from Hawaii to Massachusetts. My patient found, recognized, and picked up her friend's engagement ring from the sand on Waikiki Beach but didn't hear confirmation that it was her friend's ring until she got home, which is when she also learned that it had been lost.

Over the years I've noticed that many of my patients with bipolar disorder and schizophrenia experience a profound number of synchronicities when their illness is in its active phase, especially if they are sleep-deprived. During the times of increased synchronicities, my patients also report more psychic experiences. Both synchronicities and psychic phenomena are psychological windows into the universe's interconnectivity, and they occur together under the same psychological and physiological circumstances.

One patient of mine had a very complex paranoid delusion that the American government was spending millions of dollars tracking her every move. Her delusion, which occurred because of a conversation she overheard while in line at the post office, didn't make any sense to anyone but her. She repeatedly found herself behind cars with vanity license plates that fit perfectly into her delusional thinking. She photographed the plates while driving and arrived at my office with a pile of photos one day. She hoped that the license plate photos would prove her delusion. I never doubted that the license plates existed because my experience had been that my patients' synchronicities are almost always real events. However, I often differ with patients about how to interpret the synchronicities.

Synchronicities increase exponentially during states of consciousness in which our limbic systems are highly activated. The limbic system assigns emotion and meaning to people, animals, numbers, colors, events, and so on. So when our limbic system is more active than usual, so are our temporal lobes, and the world around us becomes more meaningful and symbolic. If we are paranoid, the symbols are threatening, but if we are in a state of awe, the symbols can be mystical or spiritual. People can also experience a combination of feelings and types of symbols.

My patient's story illustrates one of the intriguing interconnections I've seen with synchronicities between our inner reality (or thoughts), and our external reality (or experiences). Periods of increased synchronicities don't appear to be merely the result of our being on the alert for the symbols that have become very meaningful to us. When our limbic system is in an activated state, our thoughts seem to increase the likelihood of experiencing these synchronicities, almost the way a magnet attracts metal. In this sense some synchronicities may be a variant of psychokinesis, which is when our thoughts influence the physical world.

MYSTICAL EXPERIENCES

Mystics have claimed to see the universe's interconnectivity for millennia. They believe that other dimensions exist because they have had direct experience of them. For example, Sri Aurobindo, a twentieth-century Indian mystic, claimed to see an extra dimension when he entered an altered state of consciousness.

People who have experienced mystical states of consciousness universally report that during the mystical state, one's sense of separateness from the physical world dissipates. The concepts of "I," "me," or "mine" lose relevancy or meaning.[2] One's sense of time is replaced by a sense of eternity. D. T. Suzuki, the scholar who brought Zen Buddhism to the West, described the mystical state as when "we look around and perceive that . . . every object is related to every other object . . . not only spatially, but temporally. . . . As a fact of pure experience, there is no space without time, no time without space; they are interpenetrating."[3] The third major component of the state is the conviction that the "true reality" has been unmasked. William James referred to this as a "noetic" quality, or the feeling that one has tapped into a deep well of authoritative knowledge or insight that is absolute truth.

James noted three other qualities of mystical experiences. One is *ineffability*, which is the inability to describe the experience adequately in words. Lao-tse, the founder of Chinese Taoism, spoke of ineffability when he said, "The Tao that can be told of is not the Absolute Tao." Another quality is *transiency*, or the tendency for the experience to fade after an hour or two despite its lingering sense of importance. There is also *passivity*, or the feeling that one cannot control the experience once it starts. Some mystical experiences arise spontaneously. Others are induced by sleep deprivation, drugs, herbs, meditation, drumming, Sufi dancing, or breathing techniques.

Andrew Newberg and the late Eugene d'Aquili of the University of Pennsylvania measured the brain activity of meditating Tibetan Buddhists.[4] Meditation requires intense concentration, which activates our prefrontal lobes, so these lobes were more active than normal during the monks' meditation. Our parietal lobes process information about space, time, and our orientation to space-time. These lobes became less active during meditation. This change correlated with the meditators' perception that their bodies were endless and intimately interwoven with everyone and everything.

Some scientists write off perceptions during mystical states as illusions created by changes in brain activity. However, research has shown that experienced meditators also have heightened psychic abilities, so they could be uncovering hidden truths when they see an interconnected universe and other dimensions. Also, these descriptions have striking parallels with the universe described by modern physicists. Several findings and theories within physics point to a highly interconnected universe. These include superstring theory, chaos theory, and quantum entanglement, or nonlocality. And these concepts shed light on how psychic phenomena are possible.

THE PHYSICS OF CONNECTIVITY

David Bohm was a plasma physicist who later collaborated with Albert Einstein. Bohm compared our perspective of reality to that of watching live videos of the same fish from various camera angles without knowing about the setup. The videos would appear to show different fish. If we looked carefully, we'd notice that all of the fish move identically and simultaneously, but in different directions. Since the fish aren't together in a school, we might interpret these as just synchronicities, or as psychic communication between the fish. These interpretations would be artifacts of our ignorance. Bohm argued that we similarly make false interpretations about the universe because we don't consciously see the deeper level of reality where everything is interconnected and unified.[5]

Evidence for interconnectivity initially came from research on the forces that hold subatomic particles together. Scientists investigated these forces by using giant accelerators to smash subatomic particles into each other at high speeds. When two objects collide at our macroscopic level, they break into pieces smaller than the original ones, but the pieces add up to the same amount of matter. When subatomic particles collide, the sum of the pieces becomes larger. That's because particles are actually created out of the energy released in the process. These newly created particles exist for less than one-millionth of a second, a time so fleeting that they are called virtual particles.

Virtual particles skirt between subatomic particles, disappearing when they become incorporated into the receiving particle. Although virtual particles seem stealth-like, their existence doesn't violate any physical laws. In actuality, virtual particles create the major force fields in the universe through these web-like interactions between particles. For example, the exchange of virtual particles within the nucleus of atoms is what keeps the nucleus together; otherwise the like-charged protons in the nucleus would repel each other and blow the nucleus apart.

Fritjof Capra, author of *The Tao of Physics*, summarized subatomic research as follows: "Quantum theory thus reveals a basic oneness of the universe. . . . As we penetrate into matter, nature does not show us any isolated 'basic building blocks,' but rather appears as a complicated web of relations between the various parts of the whole."[6] If you became confused by that, don't worry. Even physicists have been confused by their findings. The more subatomic physicists looked for fundamental particles, the weirder their observations became. One of modern physics' biggest challenges has been to find a so-called theory of everything (TOE) that would explain these findings, as well as unite quantum mechanics with Einstein's theory of general relativity. So far superstring theories (SSTs) are the closest contenders for the TOE.

I say theories because there is more than one superstring theory, but they are lumped together because they share the basic premise that the fundamental building blocks are tiny vibrating strings and not particles. The word *string* should not be taken literally. It is an analogy taken from the physical world that we can relate to. The strings in SSTs are .0000000000000000000000000000000001 yards long. That means they are more than 100 billion billion times smaller than an electron, the negatively charged particle that orbits around the central nucleus of an atom. The ends of a string can connect to each other, in which case the string takes the form of a loop. Strings are constantly vibrating, and an infinite variety of vibration patterns can travel around their loops like the wave that travels along the circular section of a cowboy's lasso.

SST was such a radical idea at its inception that the period between 1984 and 1986 is called the "first superstring revolution." Physicists from all over the world wrote thousands of papers on it during those few years alone. When SST changed the basic unit from particles to vibrating strings, there were many ramifications. The most important is it means that the universe is far more dynamic and capable of interaction and interconnection than was previously believed.

Vibration creates interconnectivity because of a physical phenomenon called resonance, which was discovered in 1665 by Christiaan Huygens, a Dutch physicist. Resonance is the ability of one vibrating object to cause another to vibrate, or to influence its vibration. Huygens discovered resonance after hours of observing clocks with pendulums. One by one he initiated the swinging of the pendulums in a room full of clocks. Although the pendulums moved out of synch and at different tempos, they eventually, and automatically, fell into synchrony with the first pendulum that Huygens had made swing. This happened despite the pendulums' containment inside boxes.[7] Huygens's discovery was shocking at the time, because synchrony was thought to be something that only crickets, birds, fish, dancers, and other live beings did. Since his pendulum clocks were not alive, their ability to spontaneously synchronize with each other had to be due to a physical principle.

This principle is illustrated by a vibrating tuning fork. A tuning fork for the musical note C will cause another C tuning fork to vibrate just by being in its vicinity, even if it is at a different octave, but a B tuning fork will not be affected by it. Tuning forks for the same note have this effect upon each other because they share the same resonance frequency, which is the natural frequency at which an object will vibrate.[8] String theory says everything is vibrating at the subatomic level, which means that strings with the same vibrational pattern and frequency are in sync.[9]

A disturbance of the vibration pattern of a string in sync with another might cause the other string to shift its pattern to match it, or the second string might help the disturbed string regroup and resume its original pattern. This latter effect would be more likely when there are fewer disturbed strings than strings with the original pattern. So resonance at the level of strings may help maintain the integrity of objects and beings over time because they can serve as a template, or restorative memory.

According to SSTs, strings are the basic subatomic units that constitute everything, including genes and the physical characteristics coded for by those genes. Because of resonance, all of us are potentially responsive to others with a very similar makeup. This obviously means that resonance occurs at the subatomic level among members of the same species, since they share many of the same genes and physical characteristics. Resonance would also occur to some degree across species, since animals all share many of the same genes.

The closer we are to each other genetically, the higher our degree of subatomic resonance. Since all of the genetic material of identical twins is the same, they have an extremely high degree of resonance. But this resonance may not just come from their identical genes. Unless their "junk DNA" underwent mutations after separation of the fertilized egg, it would also be identical.

"Junk DNA" actually comprises 98 percent of the inherited DNA in our chromosomes, but it isn't considered part of the genome because it doesn't have genes. Its function is unknown, but it could provide a large capacity for resonance between us and our close relatives. In other words, it can contain mutations that are inconsequential as far as disease is concerned, but shared mutations may contribute to overall resonance.

The influence of like upon like in resonance requires more than just sharing the same resonance frequency. It also requires a medium that connects them. Vibration creates a wave, but only if the wave has a medium for transmission. Air is the medium for sound waves that allows the transfer of vibration from one tuning fork to another. You can feel the tuning fork's effect on air by placing your hand near the fork when it is vibrating. Another example is the ripples in a pond, which need water to propagate their waves.

But over long distances air has too much resistance or drag to act as a possible medium between objects. Resistance causes signals to dissipate, which is one reason cell phone technology requires a network of towers to provide continuous service. So what medium could allow people to be in resonance with each other at long distances? The most likely answer is that it is the same as light's medium.

Light is an electromagnetic wave, so scientists expected starlight to have a medium that allowed its transmission through space. This theoretical medium was called "ether." In 1887 the Mitchelson-Morley experiment was done to detect ether, and its results were shocking. Scientists found no resistance or friction in space. Because resistance was a characteristic of all the known media for wave transmission, the experimental results were interpreted as proof that ether didn't exist and that space was a vacuum, or complete nothingness.

That interpretation of space changed when scientists

investigated what would happen if they took away all the matter from a defined space and brought its temperature down to absolute zero, which is the lowest possible energy state.[10] What remained was called the zero point field (ZPF), which was a concept first proposed by Albert Einstein and Otto Stern in 1913. Because of the Mitchelson-Morley experiment, scientists had expected the ZPF to be a vacuum, but to their surprise they discovered in 1948 that it contained a tremendous amount of energy. The energy was so great that the famous physicist Richard Feynman said, "The energy in a single cubic meter of space is enough to boil all of the oceans of the world."[11]

The ZPF became equated with the original concept of ether as the ever-present field that forms the backdrop of the universe. In this context, scientists are using "field" to mean a matrix that connects everything. The concept of fields originated with James Clerk Maxwell and Michael Faraday, who defined a field as a disturbance or condition in space that has the potential of producing a force. In other words, space can contain information that organizes, or directs, whatever interacts with it. The ZPF is everywhere and in everything because our universe is primarily space. Although the objects in our universe appear to be solid, they are far from being solid. Empty space is 99.999999999 percent of an atom, so the nucleus of an atom takes up as much room in the atom as an ant on a football field.

The ZPF is like a vacuum only in that it has no resistance or friction. This means that waves transmitted in the ZPF don't fade and disappear the way ripples in pond water do. ZPF is an ideal medium for light, but it is also ideal for resonance between strings since there is no distortion in its signals. The vibration patterns of strings have the potential to go on perpetually because the ZPF doesn't have the friction to stop them.

DAVID BOHM

Other evidence that the universe is interconnected comes from the work of David Bohm, who was working on plasma at the Radiation Laboratory in Berkeley, California, in 1943. Plasma is a phase of matter entirely distinct from solids, liquids, and gases. It is a state in which negatively charged electrons have been stripped from their atoms. We rarely interact with it, but plasma is actually the majority of visible matter; it is used for flat panel video displays (plasma screens), but is also what many stars are made of.

When electrons are no longer attached to atoms, they behave very differently. Electrons normally repel each other because they are negatively charged; like charges repel and opposite charges attract. But Bohm saw the plasma electrons join each other and act collectively rather than as individuals. Plasma is sometimes called an ionized gas, which differs from an ordinary gas. Atoms in normal gases move about randomly, whereas plasma electrons behave in an orderly fashion, as if they are part of an interconnected whole. Bohm's observations made him comment that it was as though the electrons were alive. He concluded that something other than the four known physical forces was determining the precise path they followed. In physics the four forces are electromagnetism, gravity, and the weak and strong forces involved in the integrity of an atom's structure.

In 1959 Bohm and a research assistant, Yakir Aharonov, observed the movement of plasma electrons under the influence of magnetic fields. Surprisingly, the electrons behaved differently depending upon whether there was a nearby magnetic field, even when the electrons were in a place where the magnetic field's strength was zero and shouldn't have affected them. Somehow the electrons received information about the magnetic field. That led Bohm to propose that an unknown, fifth force field provided information about the whole environment. He defined it as a field that exists at the deepest level, pervades all of space, and is equally powerful everywhere (unlike gravitational and magnetic fields, which weaken with distance). His description of this force field sounds like it may be the same as the ZPF, or at least a component of the ZPF.

Bohm also compared the universe to a giant, flowing hologram, constantly moving and evolving. Holograms are images created artificially by laser technology, and the characteristic important to the analogy is that each part of a holographic image contains information about the whole. So if you cut a piece of film with a holographic image, each of the pieces will have an exact, but smaller, copy of the original image. Similarly, if you break a glass cube that contains a three-dimensional holographic image, each of the shards of glass will have the complete image.

Bohm proposed that there may be an infinite series of hierarchical forces within the universe, just as a hologram has images embedded within images. He divided the universe into an "explicate order," which corresponds to the world we experience, and a deeper, hidden "implicate order." The "implicate order" is where everything arises and eventually returns, like the virtual particles that flit in and out of existence.

MULTIDIMENSIONALITY AND THE BIG BANG

The mathematical formulas of superstring theories predict that there are many more dimensions than the four we perceive. Depending upon the theory, the formulas predict ten, eleven, or twenty-six dimensions. The capacity for interconnection is magnified by the addition of these extra dimensions.[12] The way in which extra dimensions increase connectivity can be understood by looking at what happens when we go from two to three dimensions. I learned about this as a child when I read Edwin Abbott's classic 1884 novel *Flatland*, which is set in a world of two dimensions. The story is about a man who tries to teach his granddaughter about a third dimension. It is now available as an animated movie.

The difference between a three-dimensional world and a two-dimensional one can be seen in maps of the earth that are drawn in the shape of a ball that was sliced and flattened, which is the only way to prevent distortion of the size and shape of countries when representing a sphere in two dimensions. The extreme left and right ends of the map are the same geographical place. On a three-dimensional globe these ends overlap and you can see that they are the same place. By going from two to three dimensions, these places no longer appear to be on opposite ends of the earth. If the earth were flat like these maps, we would have to travel east for more than 24,000 miles to arrive at a place that is only one inch west of us on a sphere. It is hard to imagine what ten-, eleven-, or twenty-six-dimensional worlds would look like. But the significance of these additional dimensions is that many of the things that appear separate to us in three spatial dimen-

Figure 4. The Goode projection map, an example of a two-dimensional depiction of a sphere.

sions may actually overlap. They just appear separate because of our limited ability to perceive dimensions.

The universe's interconnectivity is a function of its origin, which is thought to have occurred around 13.7 billion years ago. The Big Bang, still science's most popular theory for the universe's origin, states that the entire universe was condensed to the size of a grain of rice before the Big Bang. Recent work suggests that the Big Bang may actually have been a "Big Bloom," or an unfolding rather than an explosion, but in either case it dispersed everything without destroying its connectivity.[13]

All of the universe's matter was synthesized in the first milliseconds after the Big Bang, but only 4 percent of energy/matter in the universe is visible to our eyes, aided or unaided. The rest of energy/matter is "dark matter" and "dark energy." Dark matter is thought to make up 23 percent of all matter/energy, whereas the estimate for "dark energy" is 73 percent. Dark matter does not radiate light and is therefore invisible, but scientists are aware of its existence because of its gravitational effects in outer space. Dark energy has been proposed because there appears to be an antigravitational force causing the universe to expand rather than contract. The importance of dark energy and matter is that scientists still do not understand what they are, yet they comprise the vast majority of matter. Perhaps they are part of the matrix that interconnects everything, additional evidence for hidden dimensions, or evidence that this matrix contains unknown forces.

NONLOCALITY

An important concept from physics that has been used to explain psychic phenomena is nonlocality. Nonlocality refers to a particular connection between two particles that are entangled or coupled, usually because they were created by the same reaction. Because of this nonlocal connection, an influence on one will be experienced simultaneously by the other even after they have been separated by vast distances.

Einstein could never accept the existence of nonlocal connections and devised a thought experiment known as the Einstein-Podolsky-Rosen (EPR) experiment in 1935 to disprove it. Like all of Einstein's thought experiments, the EPR experiment consisted of logic and mathematical proofs. Unfortunately for Einstein, the results were the opposite from what he expected and were consistent with nonlocality.

A simplified version of the thought experiment is as follows: Electrons spin around their axis in one of two directions, up or down. When two electrons originate from a common source they are "entangled," and one essential characteristic of entanglement is that the electrons' spins will always maintain opposite directions from each other, even when the electrons drift far apart. This continued relationship between their spins is because the two-particle system operates as an indivisible whole. If the spin of one electron is altered, the other has to automatically and instantaneously change. The change of the other electron's spin cannot be the result of a signal from the altered electron because the change occurs instantaneously, requiring the signal to travel faster than light. Since Einstein had proved that traveling faster than light was impossible, the only possible explanation for the instantaneous change would be their entanglement, or nonlocal connection.

Nonlocality remained a theoretical concept until 1982, when Alain Aspect's research team in Paris successfully proved it with photons. Aspect's work has been validated by others.[14] Experiments by Zhi Zhao and colleagues were able to prove quantum entanglement of several photons.[15] Zhao's experimental results were reported in 2004 in *Nature*, which caused excitement in the press because entanglement could provide a mechanism for teleportation.[16] Nonlocality may sound strange, but it is perfectly consistent with the type of interconnection possible through multiple extra dimensions. In other words, like the edges of the two-dimensional map, entangled electrons or photons may still overlap even though they look separate in our three visible spatial dimensions.

MAKING ORDER OUT OF CHAOS

Chaos theory derives its name from the fact that the systems it describes appear disordered. However, the theory shows that these chaotic systems actually have an underlying order. Douglas Hofstadter, Ph.D., said, "It turns out that an eerie type of chaos can lurk just behind a façade of order, and yet, deep inside the chaos lurks an even eerier type of order."[17] The theory was developed by Edward Lorenz and expanded upon by others. Lorenz was a meteorologist who was trying to predict weather in 1961 with a computer. His model for weather used twelve equations. In order to save time, he rounded a number down from .506127 to .506 and entered it into the computer. The mathematical sequence that came out was dramatically different from the one that used the unabbreviated number.

Measurements in most science experiments are extremely difficult to obtain with accuracy to three decimal places; therefore to have this minor change profoundly affect the data had disturbing implications. This significant effect caused by minor changes in the initial conditions became known as the "butterfly effect" because the numerical difference was expected to be as inconsequential to the mathematical sequence as a butterfly flapping its wings would be to weather.[18] The effect was so radical that it was as if that butterfly caused a tornado.

The second component of chaos theory came after Lorenz plotted data from an experiment where the results looked entirely random before being plotted. The plotted graph showed a double spiral that maintained the same shape, but without repeating the same dimensions. The graph's appearance resembled a nautilus shell or a pinecone, both of which are logarithmic spirals. Another form of order found in nature are fractals, in which the same shape is embedded within itself over and over again. An example is the fern branch whose leaflets show the same shape as the branch. Holograms are another example, but their embedded nature doesn't become apparent until they are broken. But despite ubiquitous examples in nature of fractals and logarithmic spirals, only two kinds of order were recognized previously. One

Figure 5. A fern branch.

was when nothing changed and the other was when something repeated itself exactly. The plotted graph of a double spiral was neither of those, but it was still ordered.[19]

What initially appeared to be chaotic to Lorenz was actually a system that reflects back and forth upon itself, similar to mirrors that face each other and reproduce an image over and over again in increasingly smaller versions that are embedded in each other. Reflections back and forth create a feedback loop, which can greatly magnify any distortion or disturbance in the system, hence the butterfly effect.

These feedback loops are another demonstration of how interconnected the universe is. That the universe runs as smoothly as it does with this much feedback and interconnectivity says that there must be more order in the universe than is generally acknowledged. This is because the feedback that can create chaos can also create resonance and synchronization.

FEEDBACK AND SYNCHRONY

It is adaptive for systems and living creatures to synchronize and work collectively, which is why nature has mechanisms at many levels by which this can occur. Feedback enables this to happen and occurs at all levels, from the subatomic level of strings to complex societies of social beings. On an intermediate level, our brains and hormonal systems contain multiple feedback systems to keep us from getting out of balance. Most of the time our body's physiological feedback systems achieve homeostasis, or balance.

Since human societies are so complex, feedback systems in them are more difficult to observe than in insect societies. A fascinating example of feedback and synchrony in an insect's social system is one that has been noted for hundreds of years by Western travelers to Southeast Asia. There are regions where enormous congregations of fireflies blink on and off in unison. Their spectacular displays can stretch for miles along the riverbanks.[20]

The biologist John Buck traveled to Thailand in the 1960s and kept scores of these fireflies in captivity. Initially they blinked off and on incoherently, but after a short time an increasing number began to blink in unison. Buck discovered that the blinking was regulated by an internal oscillator that could be reset by feedback from the other fireflies' light. Each species of firefly has its own characteristic pattern. For mating to take place the female firefly must respond to the male's flashes of light at exactly the same time interval as his flashes. Fireflies in other regions don't do this, which is why the phenomenon was so intriguing. Scientists have suggested that these fireflies had more success in attracting mates from farther distances by blinking in synchrony, which is why this behavior has spread and been passed down for hundreds of generations.

Our brains are oscillators that change their frequency depending upon our state of consciousness. And humans can synchronize their EEGs with each other, despite being in separate rooms. As we've seen, when people have been asked to be in psychic contact with one another, changes in the sender's EEG show up in the receiver's EEG. These studies show that, like the fireflies, our brains can receive and respond to signals from other members of our species.

The fireflies use light, which is an electromagnetic signal. If the feedback signal between humans is electromagnetic, it would have to be of a very low frequency because higher electromagnetic frequencies are blocked by the Faraday cages in which these experiments were done. What is important is that despite the vast individual differences in how our brains are wired, our brainwaves can become synchronized with those of someone else.

This suggests that our brains share common underlying resonance frequencies, but we have control over whether we experience synchronization with another by what we focus upon.

Synchrony is easier when you share the same gene pool, as is the case among fireflies and identical twins. For humans, being in sync also becomes more likely when our brains are engaged in particular patterns of brain activity, such as meditation. It is also easier when we form an emotional bond with the person we want to be in synchrony with. Emotional bonds involve the limbic system, and in the book *A General Theory of Love*, authors Thomas Lewis, Fari Amini, and Richard Lannon explain the concept of "limbic resonance." This is the ability of most mammals to become attuned to the inner states of others. Limbic resonance is highly activated when a mother and her child, or two lovers, gaze into each other's eyes and feel love for each other. This activity enhances their psychic attunement. For example, lovers often find themselves thinking the same thoughts.

We all have the capacity for more psychic connection by enhancing our social bonds and engaging in practices that synchronize our brain waves. Research has shown that it helps to both synchronize the activity between our left and right cerebral hemispheres, and between our brains and those of other individuals. It is a matter of engaging in activities that reinforce the feedback loops for this. But many of us engage in behaviors that create negative feedback loops, and we are out of balance internally. We also become disconnected from others. We can see from both chaos theory and our observations of human behavior that feedback loops can work for us or against us. And members of our species can walk along that fine line between order and chaos.

Chapter 11

THE ESSENCE OF TIME

It is believed by most that time passes; in actual fact, it stays where it is.

—Dogen, Zen master

Time and Space are modes by which we think and not conditions in which we live.

—Albert Einstein

The only reason for time is so that everything doesn't happen all at once.

—Albert Einstein

Dogen's statement that time "stays where it is" is the opposite of how most of us view time. But our view is a man-made construct and not necessarily reality. An alternative view of time would be required in order to believe that psychics really see into the future, rather than just try to predict it. Einstein concluded that our view of time is an illusion, but his work has been too abstract to influence the way people in modern cultures usually think of time. In fact, not only do we think of time as passing, but we've become increasingly obsessed with this idea.

Mankind's obsession is reflected in our clocks, which have become increasingly precise, so much so that many modern clocks don't allow for any human error; they are set by a wireless signal that synchronizes them. We say "time is money" and feel "pressured by the clock" because our schedules need to accommodate demands from work or children with school and extracurricular activities. Our culture is so preoccupied with aging that we spend billions of dollars each year to prevent, undo, or hide time's effects on our bodies. So how can anyone say that time doesn't really pass?

The answer will be more evident after reviewing the various models for time, how the perception of time changed, and the scientific evidence for the model provided by physicists. These all lead to the conclusion that our experience of time differs from the deeper reality, a reality that allows for precognition.

THE PERCEPTION OF TIME

Just as our brains limit our ability to perceive other dimensions and the interconnectivity of our universe, our brains also influence our perception of time. For example, we may have a dream that seemed to last hours, and indeed would have needed hours for all of the events to have occurred. However, we know from dream studies and our own experience that such dreams can take place in a matter of minutes. This is similar to future memory. During a time period that only lasts a few minutes, the person "remembers" future events that take significantly longer when they actually transpire.

Our internal sense of the passage of time can also vary when awake. In general, the busier and older we are, the faster time appears to go by. People who are highly emotional can experience an even broader variation. Similar to dreaming, heightened emotions cause an increase in our limbic system's activity, altering our perception of time.

Time perception also is altered when we come close to death. I directly observed this when I almost drowned at the age of thirteen in a canoe accident. The passage of time seemed to slow down shortly after I stopped struggling against the river's current and resigned myself to dying. At the same time my entire life passed rapidly through my mind. This same experience has been described by many others who have survived a near-fatal accident. What's intriguing about the phenomenon of a life review is that it appears to be a function of how our brains are wired. It was accidentally induced in several patients during brain surgery in the 1930s when a part of their limbic system, the hypothalamus, was stimulated by electrodes.

Intense fear can make the brain think at superspeed, in contrast with what happens externally. The contrast may be why external events appear to slow down. Charles Darwin experienced the following when he was on one of his solitary walks as a child. He often was so absorbed in thought that he didn't pay close attention to where he was going:

> I walked off and fell to the ground . . . the height was only seven or eight feet. Nevertheless, the number of thoughts which passed through my mind during this very short, but sudden and wholly unexpected fall, was astonishing, and seemed hardly compatible with what physiologists have, I believe, proved about each thought requiring quite an appreciable amount of time.[1]

P. M. H. Atwater, the author of *Future Memory*, experienced an altered sense of time perception when she was making pickle syrup, which boiled over and created a sticky mess in her kitchen. Although she says that this triggered a state in which she was calm, she also admits that she normally would have been screaming in horror. This combination of being calm while horrified can occur when we are in the midst of responding to a crisis. She described her experience as "time and space overlapping and then converging with each other." In this altered state she perceived her movements as significantly slowed down, but in reality she was able to complete what would normally be an hour's worth of cleaning in six minutes. This included scrubbing and mopping the floor three times.

The heightened activity of the limbic system and its effect on the perception of time can also be seen in bipolar disorder, which was formerly called manic-depressive illness. In bipolar disorder one's moods can go to extremes, both up and down. During the manic or up phase, one's thinking is usually sped up. One's speech is often pressured, and thoughts can come so fast that speech can't keep up. When this happens one may sound like a sped-up record that skips over words or larger sections of dialogue. Sometimes the manic person doesn't see himself as sped up, but rather complains that everyone else is slowed down. Depressed people can display the opposite. They often think, talk, and move more slowly, while perceiving everything around them as happening too fast.

A theory about our perception of time was presented by Itzhak Bentov, a biomedical engineer turned cosmologist, in *Stalking the Wild Pendulum: On the Mechanics of Consciousness.* He described two types of time and space—objective and subjective. Objective time and space are what we measure. We are generally in sync with others because of a consensus about objective time and space. Subjective time and space are products of our unconscious mind and are capable of expansion and contraction. This is why time is experienced so differently in dreams, which occur in the unconscious.

When our conscious and unconscious minds are in communication with each other, this can include a changed perception of time. Increased communication with the unconscious during mystical states leads to what Bentov calls "convergence." The mind appears to be everywhere all at once. In other words, our personal consciousness appears to fill the entire universe instantaneously. Sleep deprivation can put us into this state, which was described by Charles Lindbergh after his twenty-second hour of vigilance at the controls of *The Spirit of St. Louis*. His description makes it sound dangerous to fly planes during convergence, but Lindbergh's journey was successful.

> There is no limit to my sight—my skull is one great eye, seeing everywhere at once. . . . I'm not conscious of time's direction. . . . All sense of substance leaves, there is no longer weight to my body, no longer hardness to the stick. The feeling of flesh is gone. . . . I live in the past, present, and future, here and in different places all at once. . . . I'm flying in a plane over the Atlantic Ocean, but I am also living in years now far away.[2]

MODELS OF TIME

Man's varied experiences of time have led to different models for it. A common analogy for time in the West is that of an arrow moving forward. The tip of the arrow represents the present moment and the shaft represents the past. The future isn't part of the arrow. Instead, the future is an open sea of potentialities rather than well-defined like the rest of the arrow. Since the future is entirely open and undefined, this model doesn't allow us to psychically tap into the future for specific details of what will happen. We can only make *predictions* about the future, or about where the arrow is headed. Our predictions become increasingly inaccurate the further into the future we attempt to make them. I can make a very good guess about what my life will look like two minutes from now, but I'm much less likely to accurately predict its content in nine years. In contrast, precognition's accuracy isn't limited to the near future. Therefore, evidence for precognition suggests that the "time is an arrow" model is incorrect.

Many indigenous cultures conceive of time in a way that allows for precognition. In their model the past, present, and future all coexist on a line, albeit not a straight one. Instead, it is represented by a loop. This model is influenced by the cycles of life: the seasons, the cycles of the moon, the day-night cycle, and the birth-death-rebirth cycle. The line can also take the form of a spiral, which depicts both the cyclical nature of time and the changes over time that make each cycle different from the preceding ones. The loop model has been popular among several Native American tribes, who often describe the loop as a hoop that contains a central region of timelessness. During altered states of consciousness they can tap into the center, where they feel a sense of timelessness, or an eternal present.[3]

THE PHYSICS OF TIME

In March 1955, the year of his own death, Einstein expressed his conclusions about time in a letter to his friend's widow:

> Now he has departed from this strange world a little ahead of me. That means nothing. People like us, who believe in

> physics, know that the distinction between past and present and future is only a stubbornly persistent illusion.[4]

Although we can experience time differently, depending upon our state of consciousness, different states of consciousness in themselves aren't sufficient to allow us to discard the usual notion of time; these experiences could be explained away as illusions. A new model for time needs physics to support it, and physics does. It says that time isn't a constant, but rather varies depending upon one's frame of reference. Physics also supports the idea of timelessness, or that the past, present, and future coexist. These conclusions are the result of work that started with Einstein more than a hundred years ago. In order to understand how these conclusions were reached, I'll briefly review how the scientific concept of time evolved.

One of the first changes was the realization that time isn't a universal constant, or something that remains the same regardless of the conditions under which it is measured. In fact, much of what we think of as physical constants are only constants under certain conditions. Two examples are the speed of light, which slows down when it travels through water, and the boiling point of water, which decreases at high altitudes.

Einstein showed the mutability of time in his *gedanken*, or thought experiments, that led to his special and general theories of relativity. What we call "time" is only something that our clocks measure, and there is no independent clock outside the universe that exists as a gold standard for the passage of time.[5] Instead, our experience of time is relative to our frame of reference. One person might perceive two events as simultaneous, but from another frame of reference the events may be seconds apart. It all depends upon how fast the observer is traveling relative to the location of the two events.[6]

Another change to the concept of time was to see it as inseparable from space. In other words, space and time form a continuum which Einstein called space-time. Space-time is a matrix for our universe and is warped or curved by the massive planets and stars within it.[7]

Experiments in the physical world have supported Einstein's theories.[8] For example, Einstein predicted that light from distant stars would be deflected by the warped space-time around the sun. Anomalies in Mercury's orbit in 1915 substantiated his theory, as did observations during a solar eclipse in 1919. And then deflection of radio signals from NASA's Cassini spacecraft in 2002 provided proof of the curvature of space-time with a precision that was fifty times greater than previous measurements.

In physics, there are incompatibilities between quantum theory and Einstein's theories of relativity. Each theory has good experimental evidence and cannot be thrown out. Quantum physics describes the subatomic universe, whereas Einstein's theories address the large-scale universe, so some people don't let the incompatibilities bother them. But physics can unify these theories into a comprehensive theory if the idea of time undergoes further revision.

One attempt at unification was made by the renowned physicist John Wheeler and the late Bryce DeWitt. They developed the Wheeler-DeWitt equation, which works only if we give up the notion of time and accept the universe as timeless. In other words, our concept that time exists and is divisible into the past, present, and future arises from how we experience life, rather than being a reflection of a universal reality.

Superstring theories are another attempt to unify physics. SSTs allow the existence of tachyons, which are particles that move faster than light. Whether or not tachyons are real is important because this part of SSTs would force us to change our concept of time. Tachyons were thought not to exist because of the implications of the equation called the Einstein-Lorentz transformation, which is usually abbreviated as $E = mc^2$. This equation says that the mass of something increases exponentially as it accelerates toward the speed of light. The object's mass would become infinite at the speed of light, but that would be impossible because infinity is defined as an unlimited number that is too large to ever be achieved. Also, the greater the mass of an object, the harder it is to accelerate to a higher speed, so as an object gains speed and mass, it becomes increasingly difficult for it to reach, let alone exceed, the speed of light. Photons can travel at the speed of light because they don't have any mass; consequently the equation doesn't affect them. Therefore, the reason Einstein said that nothing can travel faster than light relates to a "singularity," which is a boundary where the laws of physics break down.

Thus far there is no clear experimental evidence for tachyons, although many physicists believe in them based upon the implications of equations that are attempts to unify physics. Tachyons require physicists to use "imaginary" numbers in their equations, which creates ramifications for time.[9] Stephen Hawking addressed the implications in his book *A Brief History of Time.* He said that if real numbers are replaced by imaginary numbers, one can go backward and forward in time depending upon the circumstances.

You might think this is nonsense because of the term "imaginary numbers," but that term does not mean that imaginary numbers have no relevance to reality. Mathematics was created by man to express and predict universal laws, and imaginary numbers are as real as "real" numbers. They are also necessary to other areas in physics.

Hawking called the time created by using imaginary numbers "imaginary time." He also said:

> This might suggest that the so-called imaginary time is real time, and that what we call time is a figment of our imaginations. In real time, the universe has a beginning and an end at singularities that form a boundary to space-time and at which the laws of science break down. But in imaginary time, there are no singularities or boundaries. So, maybe what we call imaginary time is really more basic, and what we call real is just an idea that we invent to help us describe what we think the universe is really like.[10]

A theory that could do away with singularities, such as the ones associated with the speed of light, has appeal. It is why Hawking says that imaginary time may be closer to the truth. It means that time is not an arrow; the past, present, and future all coexist; and we can theoretically move through time in either direction and at varying speeds.

PRECOGNITION AND FREE WILL

Precognitive experiences, mystics, and physics all suggest that the future already exists, at least in some form. A central question this raises is: how can we have free will if the future already exists? Another way of asking it is: does this mean our lives are scripted and we have no control over them? The answer is no. Even Atwater, whose descriptions of future memory make the future appear scripted, insists that free will definitely exists.

One way to look at this paradox is to start out with the premise that the universe is timeless. Our brains, for the most part, lock us into only experiencing the "now" as a vivid and full experience. This can be compared to seeing only one frame of a movie at a time, even though the whole movie exists. We exercise our free will during every instant that we call "now," and our lives consist of a seamless succession of "nows." So during our life we always experience our free will, but if we could look at our lives from a different frame of reference, we'd see that the future already exists.

One way of reconciling the concept of free will with an already existent future is to regard the future as still mutable, although limited by certain boundaries. We all believe that the future is influenced by the present and the past. However, there is some evidence that the influence can go in either direction. In other words, the future can influence the present. That is how people explain future memory and precognitive dream episodes in which the real outcome of the future event was favorably different. In these cases the person used information from the future to make better choices.

The idea of obtaining information from the future and using it to change the outcome sounds like the science fiction movies in which people time travel to go back to the past to undo something and prevent a future disaster, but today's science fiction often becomes tomorrow's reality. There are several accounts in which people had a precognitive dream that saved their life. But so far there are not reports of anyone going back to the past and undoing something—if there is a feedback loop within time, the only convincing evidence shows that it might occur between the present and the future.

Most of us don't think that we are able to tap into the future, but J. W. Dunne's work on precognitive dreams led him to believe that we may all have these experiences on a regular basis but not recognize them.[11] He said that this is because we have a bias toward thinking that only the past influences our present and we ignore, or don't look for, the influence of the future on the present. He believed that our dreams are a combination of images from past and future experiences.

Dunne's argument is that we usually forget our dreams if we don't write them down. We also tend to analyze them based upon what we know about the past, which makes the assumption that the dreams aren't precognitive. Unless the dream had a strong emotional impact, we might not remember or recognize a future event as having been foretold by it. Dunne kept extensive diaries with detailed entries of both his dreams and life events. After writing down the events of his day, he would look at earlier pages of his diary to see if there was any correlation to dreams that he had had in the preceding days or weeks. This made it easier for him to recognize when a dream was precognitive. He found that it was a fairly common occurrence, even if much of the material was mundane and otherwise wouldn't have caught his attention.

The future ain't what it used to be.

—American poet Lewis J. Bates

For millennia people sought the advice of prophets and mystics in order to know the future. But as we were developing better clocks during the seventeenth and eighteenth centuries, rational science disassociated itself from anything having to do with the mystical or the occult. Our scientific concept of the future has changed and allows for precognition, an important part of ancient belief systems. The science of the twentieth and twenty-first centuries is bringing us back to a view of time that is similar to that found in ancient traditions but more sophisticated. Science builds upon itself such that its concepts, including its view of time, often result from revisiting old ideas, but with a new twist.

Chapter 12

THE SUM OF THE PARTS IS GREATER THAN THE WHOLE

To see a world in a grain of sand,
And a heaven in a wild flower,
Hold infinity in the palm of your hand,
And eternity in an hour.
—WILLIAM BLAKE (1757–1827)

THIS FAMOUS SECTION of William Blake's poem predates research findings by modern physicists, yet "seeing the world in a grain of sand" evokes the idea of a holographic universe and "experiencing eternity in an hour" speaks to the relativity of time. Poets and artists commonly express concepts before they are discovered in science, suggesting their unconscious is in touch with a deeper reality.

Earlier chapters have discussed some of the physics that describes that deeper reality and what we know about the changes in brain activity that can allow us to experience it. All this information can now be integrated into a model that I call the Möbius mind. It pulls together what is known about psychic phenomena and science, and addresses what consciousness is, the relationship between our inner and outer worlds, how we access information about times and places other than the here and now, and where this information is stored.

WHAT IS CONSCIOUSNESS?

The essence of consciousness could be a form of energy, matter, or a previously unacknowledged force. According to classical physics, energy is the ability to do work and matter is the substance of which physical objects are made. Because energy and matter can be converted into each other, they are often thought of as different phases of the same thing: energy/matter. Forces, such as gravitational or magnetic forces, act upon and organize energy/matter. Consciousness acts like a force when our thoughts change our brains' wiring; psychokinesis research suggests that consciousness acts like a force because it can affect RNGs and the growth of bacteria; and consciousness also acts like a force that opposes entropy when it keeps our bodies from falling into decay or disorder. So the description of a force fits consciousness better than the description of energy or matter.

The definition of force has evolved from the Newtonian concept of the ability to cause an object to accelerate, which describes the force I exert to throw a ball. When modern physicists talk about nature's forces, they are referring to the four invisible, yet powerful, influences that underpin our very existence: gravity, electromagnetism, and the strong and weak forces in the nucleus of the atom. Each force has an associated field, which is the space around an object where the force can act on another object. For example, the gravitational field of the sun keeps the earth in orbit around it.

Evidence that consciousness acts like a force field comes from research in quantum physics, which suggested that it was no longer possible to regard the physical world as separate from our conscious impact upon it.[1] This conclusion came from an

1801 experiment by Thomas Young that studied whether light traveled as a particle, in a straight line, or as a wave, which spreads outward.

Young's famous double-slit experiment consisted of placing a light source in front of a screen that had two holes a few millimeters apart. Another screen was placed behind the first screen to capture the image of the light that passed through the two holes. The image, as expected, was of two patches of light. When the holes were made smaller, the patches of light also became smaller. Young made the holes even smaller. Then something unexpected happened. The patches of light didn't reduce in size. Instead they became larger and fainter. This was not possible if light was a particle, because particles move in straight lines. When Young made the holes even smaller, he started to see light and dark lines, which were consistent with a wave interference pattern. Afterward everyone believed that light was a wave.

Then Max Planck and Albert Einstein showed that electromagnetic energy, or light, comes in discrete packets of energy called quanta, which comes from the Latin word for "how much" and resulted in the name "quantum physics." These packets of light are called photons and act like particles. Light became seen as both a particle and a wave, a feature otherwise known as wave/particle duality.

In the double-slit experiment, photon detectors were used to measure photons that were emitted individually onto the second screen. The detectors were designed to not be in the way of the photons' paths, since tracking their paths was the purpose of the experiment. When the experiment was run with the detectors turned off, a wave interference pattern appeared on the screen (as though the photons traveled like a wave through both slits).

When the detectors were turned on, the photons were seen to travel through only one hole or slit at a time. When the detectors were turned off again and everything else was the same, the interference pattern returned. The implications of the experiment were profound: the difference in how light acted depended upon whether it was observed by the detectors. It meant that light is always both a particle and a wave, but we perceive it as either a particle or wave depending upon how we observe it.

A recent experiment by Dean Radin at the Institute of Noetic Sciences added an interesting twist to Young's experiments.[2] A light beam was set up in a chamber while experienced meditators were isolated in a separate chamber. The meditators were asked to imagine focusing their attention on the distant light beam. They essentially functioned as human photon (light particle) detectors. The patterns formed by the light beam when the meditators were psychically observing it were compared with those when the meditators were not observing it. The results of nine sessions showed the influence of remote observation to be similar to the photon detectors in Young's experiment. Light acted like a particle when the meditators were observing remotely and like a wave when they weren't. The odds ratio was more than 100,000 to 1 that the results weren't due to chance.[3]

Because of observation's effect, the physicist John Wheeler proposed that we no longer use the word *observer* to describe our consciousness's role in the universe.[4] Instead, he uses the word *participator* because we cannot observe something in the quantum world without altering it.[5]

Consciousness also acts like a force in the mind-body interface. We experience our consciousness as an "I" who decides things, but our consciousness needs to interface with the necessary brain circuitry to get our bodies to act upon our ideas. Consciousness intervenes in the instant between my deciding that I want water and the initiation of my arm and hand movements to pick up a glass of water. Similarly, consciousness enables the monkey with the robotic arm, discussed in chapter 6, to initiate its movement to get food.

It's true that consciousness fields are far more complex than the fields associated with the other forces, but our bodies and brains are more complex than the magnets, massive objects, or subatomic particles involved with the other forces. Human consciousness is able to reflect upon itself and exhibit free will, which makes it different from the other forces. But although these characteristics are unique to consciousness, the differences do not mean that consciousness is not a force. All of the forces are unique. Think about the difference between electromagnetic and gravitational fields. Electromagnetic fields can carry very complex information, such as a rerun of your favorite television show or a conversation on your cell phone with your best friend, whereas gravitational fields do little more than pull smaller objects toward a much larger one.

Each of the force fields is thought to have a virtual particle associated with it that creates the field by temporarily popping into existence and transferring energy and momentum from one subatomic particle to another, like the transient push of flippers in a pinball machine that propels the ball toward a desired goal. Photons are the virtual particles associated with electromagnetism. Gravitons are thought to be responsible for the gravitational field, although no one has detected one as of yet. But some physicists believe that gravity may be different from the other forces and not have a virtual particle. Gravity might just be a function of the curvature of space-time. So if consciousness is a fifth force, it also might not have virtual particles associated with it. However, the ZPF is a soup of virtual particles, and someday one or more of its virtual particles may be found to be involved in creating consciousness fields.

In the first chapter I described the debate between the dualists and monists in regards to consciousness and the brain. I believe that the dualists are correct in that the brain is made up of energy/matter, whereas consciousness appears to be neither energy nor matter but a field. However, at the quantum level the distinctions between energy, matter, and fields begin to disappear. At that level, energy and matter are two sides of the same coin and can be converted back and forth. Subatomic particles reach a point where they can't be broken down any further; instead their collisions create other particles, energy, and virtual particles. And the virtual particles create energy and fields as they flip in and out of existence.

Superstring theorists find that the best way to explain these observations is to conceive of the subatomic world as consisting of strings vibrating at various frequencies. The subatomic world in SST is a relational, interconnected web. An understanding based upon either/or thinking no longer applies. In other words, the distinctions between the material world and consciousness become blurred. Everything may ultimately consist of the same stuff. This last statement is a form of neutral monism. So both monists and dualists may be correct. It all depends upon the level at which you look at it.

THE MÖBIUS MIND: THE RELATIONSHIP BETWEEN CONSCIOUSNESS AND THE PHYSICAL WORLD

Lama Anagarika Govinda, a Tantric Buddhist, said, "The Buddhist does not believe in an independent or separately existing external world, into whose dynamic forces he could insert himself. The external world and his inner world are for him only two sides of the same fabric." In other words, there is no objective, or independent, external reality, but rather a dynamic interplay between our inner psychological world and the external physical world. They are engaged in the act of co-creation.

This describes what I call the Möbius mind. I chose the name in reference to the Möbius strip, which is created from a strip of paper by adding a single twist before connecting its ends. The addition of this simple twist transforms the resulting loop from one that has both an inner and outer surface to one where the surfaces are one and the same. You can make a Möbius strip and see for yourself that the surfaces are contiguous by drawing a line along the center of the paper. You'll see that you end up where you began and that the entire surface area of the loop is covered by the line.

The Möbius strip is a perfect analogy for the relationship between our inner and outer worlds because we see the external world as separate from our inner world, but they are interdependent. Our brains certainly affect how we perceive the world; I can't see ultraviolet or hear a dog's whistle, but I can hear a song on the radio and read a book. In turn, the physical world shapes our brains' circuitry. Research on cats has shown that they can't figure out how to travel between vertical poles if, during a critical stage of development, they were raised in an environment

Figure 6. A Möbius strip.

where there were only horizontal poles. Their brains became wired to only perceive and interact with what they were raised with.

Our state of consciousness also affects how we interface with the world. If I'm in the state of "flow," I might play a perfect round of tennis and move through life almost effortlessly that day. In another state of consciousness I might get into a fender-bender in the parking lot, trip over a defect in the sidewalk, and spill coffee on my blouse. The relationship between my state of consciousness and how life flows is another part of what I mean by Möbius mind.

Also, our brains are part of the physical world, and our consciousness and brains mutually influence each other. Research on cognitive therapy shows that working on changing our thought patterns can rewire our brains, and neurosurgical research shows that changing our brains' wiring will alter our thought patterns. This feedback loop shows both the power of consciousness to change our brains and the way our experience of consciousness can be shaped by physical changes in our brains.

The most important relationship between consciousness and the physical world is that the physical world needs to be represented in our consciousness for us to interact with it. And psychic phenomena suggest that our consciousness field contains a far larger representation of the world than we ever imagined; they suggest that all of space and time is represented in our personal consciousness field, even though most of it is unconscious.

THE COLLECTIVE UNCONSCIOUS

Carl Jung's concept of the collective unconscious came from recognition that our unconscious contains more than just our personal unconscious. Jung saw that even when there hasn't been opportunity for cultural transmission, there are archetypes or themes that are expressed universally in dreams, the arts, and daily life. Jung believed that these archetypes are part of our heritage, just as ducks inherit the instinct to fly in a V formation. He did not believe that archetypes were transmitted by genes; instead he believed that we unconsciously tap into a collective unconscious, which C. George Boeree, Ph.D., explains as "the reservoir of our experiences as a species, a kind of knowledge we are all born with. And yet we can never be directly conscious of it. It influences all of our experiences and behaviors, most especially the emotional ones, but we only know about it indirectly, by looking at those influences."[6]

Jung's concept of a collective unconscious is important because, like psychic phenomena, it suggests that we can access information from another time or place. Jung said that we tap into the collective past of our species, but he didn't provide a mechanism for us to do this. The resonance discussed in chapter 10 provides a means for genetically similar members of the same species to be interconnected, and the chapter on time explained that information from the past can be accessed because it coexists with the present. So the collective unconscious suggests that we are unconsciously in resonance with members of our species from the past.

However, Jung's archetypes may simply be recurrent themes that reflect human nature.[7] Our choice of symbols may just reflect the way our brains are wired. Linguists have discovered that people's assignment of nonsensical words to objects follows certain patterns. In other words, even though there have been hundreds of languages, they share certain similarities and didn't develop entirely randomly. The way a word looks or sounds will evoke one type of object over another. For example, a soft, round object is more likely to be assigned a name like "bubble" with a soft sound and round letters, rather than a harsh-sounding name with spiky letters like "brittle."

Because of the complexity of our brains, scientists have thought that most of Jung's archetypes can be explained by the human brain's configuration. But when we look at species with tiny brains, we see more compelling evidence for a "reservoir for the experiences of a species." The scientific mystery of the homing or migratory abilities of various animals implies such a reservoir.

A good example is the migration of monarch butterflies, since it takes more than one butterfly generation to do the complete migration. Monarchs born in Canada near the Great Lakes consistently travel south two thousand miles in order to winter in Mexico. They die on their way back in Texas or another southern state. Their offspring continue the journey back to the original location in Canada. Over the course of a year the butterflies produce three to five generations, and the last one flies to Mexico in the winter. Although no single monarch ever completes the entire cycle, the migration route varies little from season to season. In fact, I've seen thousands of monarch butterflies go to the same exact grove of eucalyptus trees year after year.

By itself the genetic code is not sufficient to explain the monarchs' behavior. Neither is their brains' wiring. Like the uncanny similarities between identical twins raised apart, the monarchs' behavior may be due to resonance. The monarchs may tap into their collective unconscious by being in resonance with monarchs from preceding generations.

The idea that information from the past can be obtained by resonance between genetically similar beings may provide an explanation for the bizarre anecdotal reports of heart transplant recipients. Some recipients discovered postoperative changes in their reactions to things that were later found to be characteristic of the heart donors. Paul Pearsall, a psychoneuroimmunologist who studies the brain and its connections to our immune system, collected seventy-three cases and published them in *The Heart's Code*.

One of the most famous postoperative accounts was published by Claire Sylvia in *A Change of Heart*. She developed a craving for chicken nuggets six weeks after her heart and lung transplant, while having dreams about a man named Tim. The identities of donors are kept confidential, but Claire wanted to investigate the source of her dreams. She shared her experiences with a psychic who helped her locate Tim's obituary. Claire was able to contact Tim's parents and found out that he was her donor. She also learned that her donor was a big consumer of chicken nuggets and actually died with some in his pocket.

One way to make sense of these dramatic transplant accounts is that the donated organs were still in resonance with the deceased donors. Like the monarch butterflies, it suggests an ability to be in resonance with a genetically related being from the past. The organs appear to either retain or be in connection with information about the donor, which then impacts the behavior of a sensitive recipient. These accounts also imply that our brains aren't the only part of our bodies that can tap into this reservoir of personal experiences. They suggest the possibility of the body being a living hologram.

HOLOGRAPHY AND LIVING HOLOGRAMS

In a holographic universe, information about everything in space-time would be embedded in every location in space-time. (As mentioned earlier, if you break anything that contains a holographic three-dimensional image, a smaller, exact replica of the larger image is visible in each part.) This could help explain the other mystery of the monarchs' migration: their route is so specific that it is as though they navigate by a map. Similarly, when psychics see remote locations based upon geophysical coordinates, where is the map? The map can exist everywhere, if this is a holographic universe.

David Bohm believed that the universe is organized like a hologram, but that it is dynamic and not static like the holographic images we create by lasers. Bohm based his theory in part on his observation that electrons in the plasma state receive and respond to information about their whole environment. For example, the electrons' movement was influenced by a magnetic field even when they weren't physically within it.

Evidence for a holographic universe also comes from examples in nature where information about the whole is available to its parts, such as in the creation of fractals, which are complex geometric shapes that can be divided so that every part looks like a smaller version of the whole. Snowflakes are just one example.

Each of our cells has the same genetic information about our whole body, but the genes in these cells are selectively turned on or off during development depending upon the type of cell. This selectivity of gene expression enables the various cell types to look different from each other and to perform different functions. However, the shared genetics among our cells enables them to be in resonance with each other. This means that we are living holograms; each of our cells has information about our whole being.

The holographic model may also explain a major mystery in developmental biology: what guides fertilized eggs during their development into whole organisms? Without a template for the whole organism, how would the cells "know" how many times to divide, which type of cell to differentiate into, and where to migrate? And there isn't much time for information transfer among cells during the process, which is rapid and complex. After just fifty cellular divisions, humans have more cells than the number of stars in the Milky Way, and each cell undergoes an average of a hundred thousand chemical reactions per second during this early developmental process. Genes get turned on and off, but what "oversees the project" and orchestrates the actions of the genes?

Rupert Sheldrake proposed that this process occurs under the direction of a morphogenetic field, which is a field that organizes cells during embryological development. This concept was the major theory in experimental embryology during the 1920s and 1930s. In the 1940s, the concept of morphogenetic fields appeared to be corroborated when Harold Burr, a Yale neuroanatomist, detected around the unfertilized eggs of salamanders an energy field that resembled salamanders. Although the original concept of morphogenetic fields was dropped when scientists became more focused on genes, Sheldrake has brought the concept back into circulation.

It isn't just that our cells have information about the whole body. The idea that sections of our body contain a miniature representation of the entire body has been discovered empirically, and independently, across various cultures. One example of this kind of body holography is reflexology, whose basic premise is that specific sections of the soles of our feet correspond to specific parts of our body, like our large intestine or liver. Manipulating or massaging the corresponding sections on the soles of someone's feet is supposed to heal the associated body organs. Reflexology dates back to ancient Egypt and is depicted in an Egyptian wall painting in the tomb of Ankhmahor at Saqqara, also known as the "physician's tomb."[8]

Other cultures that discovered reflexology include the Native Americans, whose medicine men manipulate and stimulate the feet as a part of their healing practice. Reflexology has had a following in Europe; one of the earliest books on reflexology was published in 1582 by two eminent European physicians, Dr. Adamus and Dr. A'tatis. The version practiced in the West today is based upon the work of Eunice D. Ingham, a physical therapist who treated hundreds of patients and published her findings in 1938 in *Stories the Feet Can Tell.* Reflexology has also gained enough respect to become integrated into mainstream health care in China, Denmark, and the United Kingdom.

Iridology is another practice based on body holography. Since the seventeenth century, medical practitioners have written about correlations between markings on the eye's iris and health. One was Dr. Ignatz von Peczely, a Hungarian who graduated from Vienna Medical College in 1867. In his child-

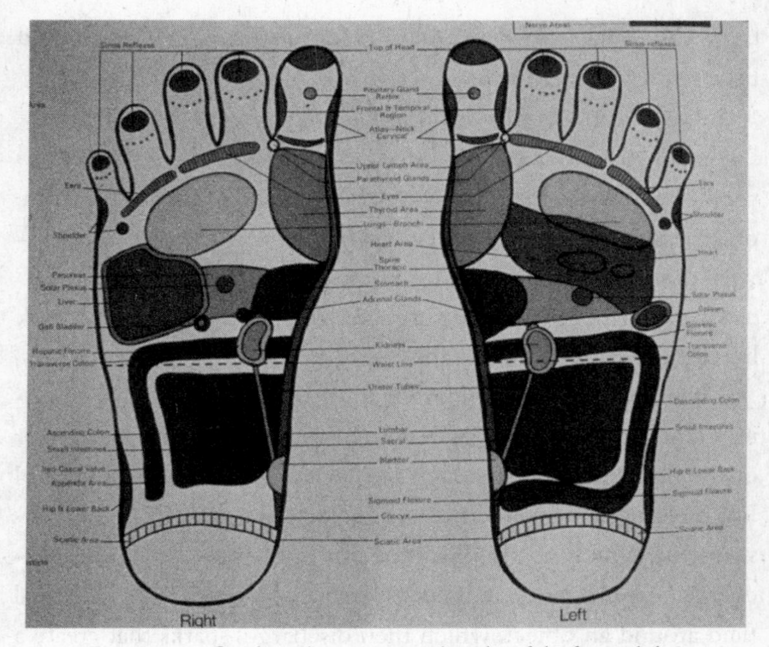

Figure 7. A reflexology chart mapping the soles of the feet and their correspondence to organs within the body.

hood, von Peczely accidentally broke the leg of an owl and noticed a black mark appear in the owl's eye. Over time the mark changed its form and shading. This intrigued him so much that he later studied the irides of patients before and after surgery.[9] He systematically recorded his correlations and published his research in the book *Discoveries in the Realms of Nature and Art of Healing*. He also established an iris chart in 1880 for diagnosis. Much of the American medical research on iridology was done by Dr. Henry Edward Lane and his student Dr. Henry Lindlahr. They made surgical and autopsy correlations and published them in Lane's 1904 book *Iridology: The Diagnosis from the Eye.* They were meticulous in their research. Lane stated that "thousands were examined before just one marking could be considered corroborated."

Acupuncture also draws upon a holographic view of the body. For example, the acupuncture points on the ear correspond to parts of the body. In fact, the chart looks like a miniature picture of a person painted on an ear. Similarly, several brain areas involved in processing physical sensations and the control of muscles were mapped by neurosurgeons, and are represented to look like little men with large hands and heads.

Other evidence for holographic organization within our bodies comes from research by Konstantin Korotkov, Ph.D., director of the St. Petersburg Research Center on Medical and Biological Engineering. He invented a gas discharge visualization (GDV) instrument, which uses a phenomenon known as the corona discharge. An electrode is used to create a high-energy electrical field around an object, which then discharges sparks that create a pattern that resembles the outer corona of the sun during an eclipse. Corona discharge patterns can be produced around inanimate objects, such as a rock, so they are not the same as auras or the energy fields in Eastern philosophy.

What is interesting about Korotkov's work is that images generated only from a person's fingertips present corona discharge patterns that provide information about the body as a whole. GDV impulses are sent to the fingertips, from which measurements are fed into a computer to generate colorful images of light around the person's body. Korotkov's research has found associations between certain aberrations in the corona discharge pattern and specific disease states. A tumor in the liver shows up as an abnormality in the liver's section of the pattern. More than four hundred GDV machines have been sold around the world, and many have been used for research supported by funding from the National Institutes of Health.

Karl Pribram, a neurosurgeon, has suggested that the brain operates like a hologram, in part because of work done on memory by Karl Lashley, an American psychologist. In the 1920s Lashley trained rats to perform tasks and then subsequently destroyed various sections of their brains' cortex. He found that the amount of cortex removed, rather than the location of the cortex, was the important factor in whether memory was affected. This violated the theory that memory is encoded in specific circuits in the brain and suggested to him that memory is stored throughout the brain's cortex. His research was published in his 1929 monograph *Brain Mechanisms and Intelligence.*

Just as our bodies have holographic features and some memory is stored holographically within the brain, our consciousness fields may be part of a hologram navigated by our brains. If this hologram contains information about the universe throughout time and space, this would explain many reports by psychics and mystics. When mystics experience an expansion of their consciousness so that it "fills all of space-time," their consciousness is probably filling up their entire consciousness fields rather than the whole universe. Remote viewing could be understood as seeing a section of the consciousness field that is normally filtered out by the brain because it isn't relevant to the present, but becomes relevant when one participates in a remote-viewing experiment. And during an OBE, one's conscious awareness could be shifting to a place within one's consciousness field where one can see things from a different angle or location. Similar to dreaming, one's perspective or orientation is no longer dominated by the sensory input from the body.

HOW DO I REMEMBER THEE? LET ME COUNT THE WAYS

Friedrich Nietzsche raised a good point when he said, "The existence of forgetting has never been proved; we only know that some things don't come to mind when we want them."[10] Is memory ever really lost, or is it just that we fail to retrieve it? The answer depends upon where and how you believe memory is stored. Memory is important because it stands at the crossroads of consciousness, the brain, and the storage of information. Several scientists, including the physicist Erwin Laszlo, propose that the storage of our personal memories may not differ from the storage mechanism of psychic information. It may just be that our brains are wired to make it easier to access personal memories than psychic information.

Neuroscience has built its model for memory based upon patients with deficits in memory and the correlations with brain damage. Scientists divide personal memory into several types. Declarative memory is what we use for factual material, such as language and what we are taught in school. It's the memory that people say we lose if we don't use. Episodic memory is the memory for events in our life, such as our first date. It stays with us if it was meaningful or emotionally charged, but the details can become distorted over time. And motor memory is the memory for coordinated actions, such as riding a bike. We can retain those skills even if years go by since we employed them.

Remembering a word depends on the cortex, where language is organized. A tiny stroke can wipe out all of the words for vegetables, tools, or some other category while leaving the words for everything else intact. In neuroscience's model, memory relies upon the strengthening of synaptic connections between brain cells in the cortex by their repetitive use. But this model has not taken into consideration people with extraordinary memories, such as savants and the synesthete known as S, who must be included for a complete model. One explanation for these individuals is that they access the information from their holographic consciousness fields.

Another challenge to science's memory model was reported by a British neurologist, John Lorber, in an article in *Science* titled "Is Your Brain Really Necessary?"[11] Lorber did a brain scan of a student with an IQ of 126 who had earned a first-class honors degree in mathematics and whose social skills were completely normal. His brain scan showed that he didn't have the usual 4.5-centimeter-thick cortex. Instead, his cortex was an extremely thin layer measuring a millimeter or so, which is 1/450 the normal thickness. Most of his brain consisted of the deep structures, such as the limbic system.

The student's condition was caused by hydrocephalus, otherwise known as "water on the brain," which occurs when the brain's cerebrospinal fluid becomes blocked, builds up in the brain, and causes brain damage by compressing the brain against the skull. The cortex takes the biggest hit because it is the outermost layer. This case was so dramatic that it led Lorber to conclude that "the cortex probably is responsible for a great deal less than people imagine." What's interesting is that the student had such a high degree of compensation, and unlike in other cases of brain injury, the compensation couldn't involve taking over other areas of cortex.

What is known about Lorber's student and the autistic savants would suggest that scientists should reconsider the idea that memory is coded by increasing the strength of specific synaptic connections among cortical cells. The student with barely any cortex doesn't prove that the cortex is not involved in memory, but it suggests that the cortex might not be essential to memory. Savants such as Kim Peek can memorize more than twelve thousand books despite having fewer connections in his cortex than most of us. When functional MRI and PET scan studies show activity in the cortex during memory retrieval, it just means that our cortex plays a role in accessing memory. And if the cortex is abnormal during critical periods of development, it doesn't become part of the search engine for memory retrieval. The research on savants and psychics strongly suggests that deeper brain structures can access information when the cortex is impaired or underactive.

The findings from clinical research on deficits in memory associated with brain damage show that the people with the most deficits are those with damage to the hippocampi, which I propose to be part of the limbic system that accesses psychic information from a holographic consciousness field. Some people sustain cortical damage after they had relied for a long time upon their cortex as their retrieval system. Because so much of their cortex is still intact, they usually don't shift to using the deep limbic system structures to retrieve memories.

So how is psychic information stored in the consciousness field? Edgar Cayce said he accessed psychic information from the "Akashic records." The word *Akashic* comes from the Sanskrit word for "space" or "ether." Cayce used that word to say that information about everything is stored in space. Erwin Laszlo equated the Akashic records with the zero point field, the "vacuum" that was found to be full of energy and virtual particles. Laszlo has argued that the brain is simply the retrieval and readout mechanism of the ZPF, which he calls the "ultimate storage medium." David Bohm and Rupert Sheldrake have similarly called the ZPF a storage device. I agree with these scientists that this information is stored in a field, which I call our personal consciousness field. The question I have is whether or not one's consciousness field is exactly the same as a personal portion of the ZPF.

The idea of memory storage in a field is not novel. Most of our modern technology for communication relies upon this capacity of fields. What is new is to regard our own personal memories, along with all of the impersonal information about everything in space-time, as stored within a field. Because our personal memories are very important to us, our brains are wired to preferentially access them. In contrast, people with autism usually become attracted to things, rather than people, in large part because people are so changeable and unpredictable to them that they are scary. So the information that autistic savants preferentially retrieve typically consists of lists of phone numbers, calendars, prime numbers, music, or something else that has order and predictability.

SEEING THROUGH THE THIRD EYE: THE PINEAL GLAND

Harold Puthoff, one of the physicists who conducted research on remote viewing, concluded that we have all of the information of the world at some level of awareness. He thought that people who were good at remote viewing were just better at damping the noise from all the other distractions. Puthoff's statements echo those by Patanjali, the ancient Hindu philosopher and author of the *Yoga Sutras*. The neurological studies on psychic abilities and the brain suggest the same thing: damping the noise involves turning down the activity of the cortex relative to the activity of the limbic system.

The secret to accessing information from our consciousness

field may boil down to a naturally occurring chemical in the brain, N,N-dimethyltryptamine (DMT). Rick Strassman, the psychiatrist who did extensive research on DMT with human subjects, was intrigued by this question: why does a potent psychedelic exist in our brains? Our brains have enkephalins as natural opiates that reduce pain, and melatonin, which induces sleep after the sun goes down, but why would brains have molecules that cause OBEs and hallucinatory experiences? DMT appears to have two adaptive purposes. One is its role in psychic phenomena, and the other is to make the process of dying less frightening when our time comes.

The source of DMT is the pineal gland, which is often called the "third eye." This refers to its rich cultural association with seeing other realms of reality. But, in at least one species, it actually has the features of a third eye. The western fence lizard (*Sceloporus occidentalis*) has an opening on the top of its skull through which one can see the upper half of the pineal organ, which has an optical lens, cornea, and retina. During the course of evolution, the pineal gland started out as an eye on top of the head that lost its ocular features and became buried. The limbic system's development buried it, but it became buried even deeper after the expansion of the outermost part of our brain, the cerebral cortex. In humans the pineal gland sits in the center of the brain, and slightly above the level of our eyes.

Dreaming and OBEs, the two major states of consciousness associated with psychic information, are both linked to DMT. Small amounts of DMT are released during dreaming, which may contribute to the vividness of dreams. DMT also facilitates psychic dreams by its effect on the limbic system. Larger amounts of DMT can cause an OBE, once again by its effect on the limbic system. So DMT temporarily breaks down our brain's filter by shifting the relative activity of our cortex and limbic system, enabling our conscious awareness greater access to information in our holographic consciousness field.

When people come close to death, the massive release of DMT can lead to a near-death experience that includes an OBE and other features, such as angels, that are comforting. Sometimes the near-death experience is very frightening, which is most likely to occur if at some level the person feels he is not living his life according to society's morals or values. Regardless of whether the NDE felt blissful or terrifying, people are usually much more spiritually inclined afterward. Also, once large amounts of DMT have been released, the brain's filtering system is often less effective. Many survivors of near death are more likely than others to experience psychic phenomena while awake. Evidence that their limbic system and temporal lobes underwent a permanent change in wiring includes the studies mentioned earlier. In one, more than 20 percent of near-death survivors showed evidence of seizures in their temporal lobes. In another, there was a characteristic disturbance of the normal sleep cycle in people after an NDE.

Mystics engage in certain rituals to expand their level of consciousness and psychic abilities. These practices appear to stimulate the brain to produce and release more DMT. During kundalini yoga some people experience a rush of energy traveling upward from the base of the spine to the brain. The kundalini experience intensifies with each successive one, eventually resulting in what some describe as an "electric storm." Many see a blinding light, which is one reason these experiences are associated with a path to "enlightenment," or a state of highly evolved consciousness with enhanced spirituality, cognition, and psychic abilities.

Engaging in kundalini yoga and fasting can activate this

process, but for some individuals kundalini experiences can happen spontaneously. For some, too, the experience can become complicated by psychotic symptoms, perhaps because of an uncontrolled release of too much DMT, or an inadequate metabolism of the DMT released. Gopi Krishna wrote extensively about his kundalini experiences, including the frightening period in which he temporarily went insane.[12] Afterward he regained his sanity and possessed greater cognitive and psychic abilities than before.

Psychics, including Edgar Cayce, often identify the pineal gland as the source of their abilities. Eastern religions also teach that the pineal gland is transformed by the kundalini experience. Scientific research supports the connection between the pineal gland and psychic abilities because the pineal gland releases DMT. Also, the pineal gland is influenced by the earth's electromagnetic fields, which have been shown by several studies to impact psychic phenomena.[13] Neither Cayce nor the Eastern mystics had modern research available to them to reach their conclusion about the pineal gland. They accessed it psychically, which makes any scientific corroboration especially intriguing.

PLACING THE MÖBIUS MIND MODEL IN PERSPECTIVE

We all sit in a circle and suppose, while the secret sits in the center and knows.

—Robert Frost

Robert Frost's statement serves as a reminder that scientific theories or models are only an attempt to approximate truth by using the facts we have available. That is the purpose of the Möbius mind model, which can be summarized as follows:

1. Psychic phenomena appear to be both real and a potential in all of us.
2. Dreaming, meditating, synesthesia, astral projection, and the minds of autistic savants are all conditions or states of consciousness in which psychic abilities can be enhanced; all show a shift away from the usual dominance of brain activity in the left hemisphere and from the cortex to the limbic system.
3. DMT is a chemical produced by our pineal glands that also exists in plants. It has been used for millennia by shamans to create psychic experiences. Consistent with the states mentioned above, it stimulates the limbic system to be more active than the cortex.
4. Consciousness has the properties of a force that can act upon the physical world both locally and remotely. The four major forces in physics have fields that influence matter within their reach. If our consciousness is a fifth physical force, it also has a field. And like other fields, consciousness would exist both within and beyond its source.
5. Occurrences of out-of-body experiences and psychic phenomena are well documented. This implies that we can access information from other locations in space and time. As a force, consciousness would not be confined to our brains, but this does not mean that consciousness travels vast distances to retrieve distant information. A simpler explanation is that we have greater access to our

entire consciousness fields during OBEs and psychic states, and that our consciousness fields contain a miniature representation of the universe. This idea is consistent with David Bohm's theory that our universe is holographic, and each part contains information about the whole.

6. Time and space cannot be separated. Instead, they form a matrix called space-time. Because the past, present, and future all coexist, foreseeing the future becomes possible. However, in this model the future can still be changed and we are able to exercise free will.
7. Our universe is multidimensional and highly interconnected. These connections and other dimensions are not readily apparent in our usual state of consciousness, during which our inner and outer worlds appear to be separate. But like a Möbius strip, the inner and outer worlds are contiguous. They also interact in a dynamic process of mutual influence.

IMPLICATIONS OF THE MÖBIUS MIND MODEL

No doubt many scientists will remain skeptical of psychic phenomena, despite credible evidence documenting its existence. But others may want to investigate psychic phenomena, intrigued by the promise that a new model for consciousness and how the mind works could make sense of seemingly inexplicable anomalies. Regardless of whether scientists validate, disprove, or refine the Möbius mind model, this new paradigm deserves serious scientific research.

One of the reasons psychic phenomena have not been more widely accepted within neuroscience may be that most neuroscientists have not investigated the existing data. Also, with the exception of a few scientists like Stuart Hameroff, mainstream neuroscientists don't attempt to integrate theories from physics into their model. Although physics informs us that the world is not as we directly perceive it through our senses, our perceptions are so compelling that they dominate our beliefs. It has not helped that many of the theories in physics have been so difficult to comprehend, even by physicists, that they have had a limited impact on what we consider to be possible. *The ESP Enigma* has presented the basics of these theories in a more accessible way to general readers and has encouraged scientists and lay readers to incorporate them in examining psychological anomalies such as precognition and telepathy.

The Möbius mind model requires a paradigm shift analogous to the Copernican revolution, which took place over the span of more than a century. Because of anomalies in planetary movements, Nicolaus Copernicus proposed in 1543 that the earth was not the center of the solar system. But the Copernican view was not accepted until Isaac Newton's work in 1687, fifty-five years after Galileo Galilei was prosecuted over his support of the concept.

Psychic phenomena and other psychological anomalies are pushing us toward developing a new theory about consciousness, although they imply a reality that is so mind-boggling that they remain controversial. But a paradigm shift is already happening, even though there is major opposition. Skepticism and opposition are characteristics of paradigm shifts, recognized by the nineteenth-century philosopher Arthur Schopenhauer, who said, "All truth goes through three stages. First it is ridiculed. Then it is violently opposed. Finally it is accepted as self-evident."[14] We have been in the second stage, but we are approaching the third.

For anyone who is interested in understanding the workings of the mind and our complex interaction with the world around us, turning a blind eye to psychic phenomena is no more an option than refusing to believe that spacecraft landed on the moon simply because we cannot explain the physics that made such a feat possible. Albert Einstein summarized the importance of exploring our ideas about ourselves when he said,

> A human being is a part of the whole, called by us the "Universe." . . . He experiences himself, his thoughts and feelings as something separated from the rest, a kind of optical delusion of his consciousness. This delusion is a kind of prison for us, restricting us to our personal desires and to affection for a few persons nearest to us. Our task must be to free ourselves from this prison by widening our circle of compassion to embrace all living creatures and the whole of nature.[15]

The ESP Enigma represents a shift that could become the next major evolutionary leap in understanding ourselves and our place in the universe. In the past we have created our own prisons by holding on to limited beliefs about what is possible. Perhaps we now can begin to open up our minds and unlock the gates.

ACKNOWLEDGMENTS

Writing this book would not have been possible without the dedicated research of many brilliant men and women, some of whom (Gary Schwartz, Dean Radin, and Marilyn Schlitz) I've gotten to know during the course of completing it. With deep appreciation I want to thank everyone whose research inspired me to write this book: my wonderful and gracious editor at Walker & Company, Jacqueline Johnson, for her belief in me, enthusiasm for this book, and gentle, insightful editing; my literary agent, Victoria Pryor, whose nurturing dedication to her authors and their books is truly remarkable; Jean Houston, a brilliant and inspirational mentor to me and others; my brother Kenneth Hennacy, a physicist with deep and insightful knowledge about consciousness; Richmond Mayo-Smith and Danny Mack for their generous support of my work; and members of my writing group who encouraged me while providing invaluable feedback (Anne Batzer, Fayegail Mandell Bisaccia, Kathy Bryon, Kathie Olesen, and Sharon Schaefer). Space does not allow me to include everyone who deserves recognition, so I extend my heartfelt thanks to all, named and unnamed, who have participated in the birthing of this book.

NOTES

Introduction

1. Francis Crick, *The Astonishing Hypothesis* (New York: Scribner, 1994), 3.
2. Adrian Parker, "We ask, does psi exist? But is this the right question and do we really want an answer anyway?" in James E. Alcock, Jean E. Burns, and Anthony Freeman (eds.), *Psi Wars: Getting to Grips with the Paranormal* (Charlottesville, VA: Imprint Academic, 2003), 111–34.
3. William James, "Essays in popular philosophy: What psychical research has accomplished," in *The Will to Believe and Other Essays in Popular Philosophy* (New York, London, and Bombay: Longman Green and Co, 1899), 319.

Chapter 1: Consciousness and the Brain

1. Physicists such as Fritz Capra, who wrote *The Tao of Physics,* have recognized parallels between these Eastern religious views and quantum physics. As a result, some physicists have adopted the mentalist perspective.

2. The pineal appealed to Descartes because of its central location, but also because it is a single structure, unlike the rest of the brain, which comes in pairs divided between the left and right hemispheres.
3. William James, "Human immortality: Two supposed objections to the doctrine," in G. Murphy and R. O. Ballou (eds.), *William James on Psychical Research* (New York: Viking, 1960), 279–308. Original work delivered as a lecture in 1898.
4. Aldous Huxley, *The Doors of Perception* (New York: Harper and Brothers, 1954).
5. The homunculi on the right side have their input to and output from the left side of the body, whereas the ones on the left side correspond to the right side of the body. All of the homunculi have a distorted shape relative to our actual bodies, because highly innervated areas such as the hands and lips require more surface area of the brain for representation than areas like our torso.
6. Positron emission tomography (PET) scans were developed in 1973. They produced color-coded pictures of the fluctuating degrees of a subject's brain activity during cognitive tasks. They require an injection of radioactively tagged chemicals such as glucose or oxygen that would be taken up preferentially by the most active cells. Now scientists often use functional magnetic resonance imaging (fMRI), which applies powerful magnetic fields that detect differences between oxygenated and deoxygenated blood, revealing differences in blood flow to regions without exposing the person to radiation.
7. Although this astronomical calculation of possible brain activity is cited often, in reality the potential combinations are fewer because the brain is organized into systems that have preferred

patterns of brain activity. For example, each of the sensory organs has its own system in the brain for processing its sensory information, and these regions form connections that are more predictable than areas that are involved in processes that allow for more flexibility and vary more between individuals.

8. Neurotransmitters set off a chain of reactions that ultimately change the expression of genes. When a gene is turned on, it produces a protein that has physiological effects. For example, researchers now believe that most antidepressants work by increasing neurotransmitters that initiate a multistep cascade that results in turning on the gene for BDNF, or brain-derived neurotrophic factor. BDNF helps brain cells survive, stimulates the creation of new brain cells, and helps create new connections between brain cells.
9. A. M. Owen, M. R. Coleman, M. Boly, et al., "Detecting awareness in the vegetative state," *Science* 313 (2006): 1402.

Chapter 2: Do You See What I See? An Examination of the Evidence for Telepathy

1. Ian Stevenson, *Telepathic Impressions* (Charlottesville, VA: University of Virginia Press, 1970).
2. Berthold Schwarz, *Parent-Child Telepathy* (New York: Garrett Publications, 1971).
3. Louisa E. Rhine, *The Invisible Picture: A Study of Psychic Experiences* (Jefferson, NC: McFarland, 1981).
4. Carl Jung believed in a collective unconscious that contained archetypal material that was inherited and universal, so his definition of the unconscious contained far more than the personal unconscious.

5. E. Gurney, F. Myers, and F. Podmore, *Phantasms of the Living* (London: Trübner Co. 1986), 1:202.
6. Hornell Hart, "Reciprocal dreams," *Proceedings of the Society for Psychical Research* 41 (1933): 234–40.
7. Raymond de Becker, *The Understanding of Dreams: And their Influence on the History of Man*, translated by Michael Heron (New York: Bell Publishing Company, 1948), 394–95.
8. Montague Ullman and Stanley Krippner, *Dream Studies and Telepathy* (New York: Parapsychology Foundation, 1970); Montague Ullman, Stanley Krippner, and Alan Vaughan, *Dream Telepathy* (Jefferson, NC: McFarland, 1989).
9. S. Krippner, C. Honorton, and M. Ullman, "A long-distance ESP dream study with the Grateful Dead," *Journal of the American Society of Psychosomatic Dentistry and Medicine* 20 (1973): 9–17.
10. I. L. Child, "Psychology and anomalous observations: The question of ESP in dreams," *American Psychologist* 40 (1985): 1219–30.
11. A. Vaughn, "A dream grows in Brooklyn," *Psychic* (Jan./Feb. 1970): 40–45.
12. S. J. Sherwood and C. A. Roe, "A review of dream ESP studies conducted since the Maimonides dream ESP studies," in James E. Alcock, Jean E. Burns, and Anthony Freeman (eds.), *Psi Wars: Getting to Grips with the Paranormal* (Charlottesville, VA: Imprint Academic, 2003), 85–110.
13. M. Bertini, H. Lewis, and H. Witkin, "Some preliminary observations with an experimental procedure for the study of hypnagogic and similar phenomena," *Archivio de Psicologia, Neurologia, Psichiatria e Psicoterapia* 25 (1964): 493–534.
14. Dean Radin, *Entangled Minds: Extrasensory Experiences in a Quantum Reality* (New York: Pocket Books, 2006).
15. Adrian Parker, "We ask, does psi exist? But is this the right

question and do we really want an answer anyway?" in James E. Alcock, Jean E. Burns, and Anthony Freeman, (eds), *Psi Wars: Getting to Grips with the Paranormal* (Charlottesville, VA: Imprint Academic, 2003), 111–34.

16. Upton Sinclair, *Mental Radio* (Charlottesville, VA: Hampton Roads Publishing, 2001).
17. Whately Carington, *Thought Transference: An Outline of Facts, Theory and Implications of Telepathy* (New York: Creative Age Press, 1946).
18. A major critic of Rhine's work has been Mark Hansel, who wrote *ESP, a Scientific Evaluation*. Hansel claimed to have effectively discredited Rhine by devising scenarios in which four of Rhine's subjects could have cheated. There was not any evidence of fraud, and Rhine took precautions against it. Nonetheless, many skeptics still discard Rhine's body of work because of Hansel's writings.
19. The institute was partially founded by the physiologist Dr. Charles Richet, who won the Nobel Prize in Physiology or Medicine in 1913 for his discovery of the anaphylaxis reaction. He was also a member of the SPR, and as the years passed devoted increasingly more of his time to the study of psychic phenomena. Five years after receiving his Nobel Prize he helped found the International Metapsychic Institute in Paris.
20. René Warcollier, *Mind to Mind* (New York: Creative Age Press, 1948).
21. Warcollier's conclusions remind me of autistic savants, some of whom reportedly have psychic abilities. The savants have a fascination with repetition, movement, and spatial relationships, but they struggle with abstract intellectual ideas and do not impose conscious imagination onto what they see. Some also have unique perceptual relationships between form and color.

22. V. Bekhterev, "Direct influence of a person upon the behavior of animals," *Journal of Parapsychology* 13 (1924): 166–76.
23. Rupert Sheldrake, *Dogs That Know When Their Owners Are Coming Home* (New York: Three Rivers Press, 1999).
24. J. Grinberg-Zylberbaum and J. Ramos, "Patterns of interhemispheric correlation during human communication" *International Journal of Neuroscience* 36 (Sept. 1987): 41–52.
25. The cages were invented in 1836 by Michael Faraday.
26. Michael Gershon, *The Second Brain: The Scientific Basis of Gut Instincts and a Groundbreaking New Understanding of Nervous Disorders of the Stomach and Intestine* (New York: HarperCollins, 1999).
27. The electrogastrogram is used clinically when someone is having problems with the stomach emptying its contents, to see if it is due to abnormal stomach motility.
28. D. Radin and M. Schlitz, "Gut feelings, intuition, and emotions: An exploratory study," *Journal of Alternative and Complementary Medicine* 11, 1 (2005): 85–91.

Chapter 3: Two Hearts Beat as One: Identical Twins and Coupled Consciousness

1. Francis Galton, "The history of twins, as a criterion of the relative powers of nature and nurture," *Journal of the Anthropological Institute of Great Britain and Ireland* 5 (1876): 391–406.
2. Guy Playfield, *Twin Telepathy: The Psychic Connection* (London: Vega, 2002).
3. Horatio Newman, *Twins and Super-Twins* (London: Hutchinson, 1942).

4. R. Sommer, H. Osmond, and L. Pancyr, "Selection of twins for ESP experimentation," *International Journal of Parapsychology* 3, 4 (1961): 55–73.
5. S. J. Blackmore and F. Chamberlain, "ESP and thought concordance in twins: A method of comparison," *Journal of the Society for Psychical Research* 59, 831 (1993): 89–96.
6. F. Barron and A. M. Mordkoff, "An attempt to relate creativity to possible extrasensory empathy as monitored by physiological arousal in identical twins," *Journal of the American Society for Psychical Research* 62, 1 (1968): 73–79.
7. E. A. Charlesworth, "Psi and the imaginary dream," *Research in Parapsychology* (1974): 85–89.
8. F.-H. Robichon, "Contribution a l'étude du phénomène télépathique avec des individus liés par la condition biologique de gemellité monozygote," *Revue française de psychotronique* 2, 1 (1989): 19–35.
9. T. D. Duane and T. Behrendt, "Extrasensory electroencephalographic induction between identical twins," *Science* 150 (1965): 367.
10. Nancy Segal, *Entwined Lives: Twins and What They Tell Us About Human Behavior* (New York: Dutton, 1999).
11. Another sad example of this happened on October 8, 1983. The mother of four-month-old twins, Samantha and Gabrielle Connolly, discovered that they had both died of sudden infant death syndrome in separate beds.
12. R. Jirtle and R. Waterland, "Transposable elements: Targets for early nutritional effects on epigenetic gene regulation," *Molecular and Cellular Biology* 23, 15 (2003): 5293–300.
13. D. K. Sokol, C. A. Moore, R. J. Rose, C. J. Williams, T. Reed, and J. C. Christian. "Intrapair differences in personality and cognitive ability among young monozygotic

twins distinguished by chorion type," *Behavior Genetics* 25, 5 (1995): 457–66.

Chapter 4: Clairvoyance: The Ability to See Remotely

1. E. Haraldsson and J. M. Houtkooper, "Psychic experiences in the Multinational Human Values Study: Who reports them?" *Journal of the American Society for Psychical Research* 85 (1991): 145–65.
2. Gina Cerminara, *Many Mansions: The Edgar Cayce Story on Reincarnation* (New York: Signet Books, 1967).
3. K. Paul Johnson, *Edgar Cayce in Context: The Readings: Truth and Fiction* (Albany: SUNY Press, 1998).
4. Edgar E. Cayce and Hugh L. Cayce, *The Outer Limits of Edgar Cayce's Power* (New York: Paraview, 2004).
5. C. Norman Shealy and Caroline Myss, *The Creation of Health* (Walpole, NH: Stillpoint Publishing, 1988).
6. C. Tart, "A psychophysiological study of out-of-the-body experiences in a selected subject," *Journal of the American Society for Psychical Research* 62, 1 (1968): 3–27.
7. Russell Targ conducted the research from 1972 until 1982, Hal Puthoff from 1972 until 1985. Ed May took over directorship of the program in 1985, though he had been on the SRI team since 1976. The last four years of the program were at Scientific Applications International Corporation (SAIC), a defense contractor.
8. Targ became involved with the Russians as a private citizen after he left SRI. He conducted two experiments between Moscow and San Francisco in 1984 with a famous Russian psychic named Djuna Davitashvili. At a distance of more

than ten thousand miles Djuna accurately described where an SRI colleague would hide two hours later in San Francisco.

9. S. James P. Spottiswoode, "Geomagnetic activity and anomalous cognition: A preliminary report of new evidence," *Subtle Energies* 1, 1 (1990): 91–102.
10. Ingo Swann, *Natural ESP* (New York: Bantam Books, 1987).
11. The latitude and longitude were often given in binary code, which is the series of 1's and 0's that can be used to represent any number. For example, 110 in binary code is the number 6.
12. Joe McMoneagle, *Remote Viewing Secrets* (Charlottesville, VA: Hampton Roads, 2000).
13. Russell Targ and Jane Katra, *Miracles of Mind: Exploring Nonlocal Consciousness and Spiritual Healing* (Novato, CA: New World Library, 1998).
14. B. Dunne and R. Jahn, "Information and uncertainty in remote perception research," *Journal of Scientific Exploration* 17, 2 (2003): 207–42.
15. H. Reed, "Intimacy and psi: An initial exploration," *Journal of the American Society for Psychical Research* 88 (1994): 327–60.
16. Russell Targ, *Limitless Mind* (Novato, CA: New World Library, 2004).

Chapter 5: The Future Is Now: Evidence for Precognition

1. Peter Hurkos, *Psychic: The Story of Peter Hurkos* (London: Arthur Barker, 1961).
2. Andrija Puharich, *Beyond Telepathy* (Garden City, NY: Doubleday, 1962).

3. Ward Hill Lamon, *Recollections of Abraham Lincoln 1847–1865*, available now from Kessinger Publishing LLC's special Legacy Reprint Series.
4. Heinz Pagels, *The Cosmic Code: Quantum Physics as the Language of Nature* (New York: Simon and Schuster, 1982).
5. John W. Dunne, *An Experiment with Time* (Charlottesville, VA: Hampton Roads, 2001).
6. John B. Priestley, *Man and Time* (London: Aldus Books, 1964).
7. S. Krippner, M. Ullman, and C. Honorton, "A precognitive dream study with a single subject," *Journal of the American Society for Psychical Research* 65 (1971): 192–203; "A second precognitive dream study with Malcolm Besant," *Journal of the American Society for Psychical Research* 66 (1972): 269–79.
8. C. Honorton and D. C. Ferrari, "Future telling: A meta-analysis of forced-choice precognition experiments, 1935–1987," *Journal of Parapsychology* 53 (1989): 281–308.
9. B. J. Dunne, R. G. Jahn, and R. D. Nelson, "Precognitive remote perception," *Princeton Engineering Anomalies Research Laboratory Report*, August 1983.
10. Russell Targ and Jane Katra, *Miracles of Mind: Exploring Nonlocal Consciousness and Spiritual Healing* (Novato, CA: New World Library, 1998).
11. Dean Radin, *Entangled Minds: Extrasensory Experiences in a Quantum Reality* (New York: Pocket Books, 2006).
12. Ibid.
13. Ibid.
14. Joseph B. Rhine, "The present outlook on the question of psi in animals," *Journal of Parapsychology* 15, 4 (1951): 230–51.

15. J. G. Craig and W. Treurniet, "Precognition in rats as a function of shock and death," in W. G. Roll, R. L. Morris, and J. D. Morris, (eds.), *Research in Parapsychology* (Metchuen, NJ: Scarecrow Press, 1973), 75–78.
16. Phyllis Atwater, *Future Memory* (Charlottesville, VA: Hampton Roads, 1999).
17. Dreams have been a major source of prophecy since ancient times. Dreams were valued so highly that kings, such as Nebuchadnezzar, used dream interpreters to guide them in affairs of the state. One of the first written records of prophetic dreams dates back to 2000 B.C. (the Egyptian papyrus of Deral-Madineh). During this era the Egyptians built temples specifically to sleep in for the induction of dreams. Theories about the origin of prophetic dreams differ according to religion. Many ancient cultures, including those of Egyptians and Jews, considered dreams to be divine revelations. The Bible has about seventy references to such dreams and visions. In contrast to viewing dreams as God-given, Hindus attributed all of them to the dreamer's soul. Ancient Vedic scripture (1500–1000 B.C.) considered telepathic and prophetic dreams to be evidence of the soul's journey outside the body to travel between worlds during dreams.

Chapter 6: Mind over Matter: Evidence for Psychokinesis

1. Brendan O'Regan and Caryle Hirshberg, *Spontaneous Remission: An Annotated Bibliography* (Sausalito, CA: Institute of Noetic Sciences, 1993); Tilden Everson and Warren Cole,

Spontaneous Regression of Cancer (Philadelphia: W. B. Saunders Company, 1966).

2. R. C. Byrd, "Positive therapeutic effects of intercessory prayer in a coronary care unit population," *Southern Medical Journal* 81 (1988): 826–29.
3. Larry Dossey, *Prayer Is Good Medicine* (New York: HarperCollins, 1996).
4. Leonid L. Vasiliev, *Experiments in Mental Suggestion,* trans. A. Gregory (London: Institute for the Study of Mental Images, 1963). It is available now through Hampton Roads.
5. William Braud, *Distant Mental Influence* (Charlottesville, VA: Hampton Roads, 2003).
6. Ibid.
7. N. Richmond, "Two series of PK tests on paramecia," *Journal of the American Society for Psychical Research* 36 (1952): 577–87.
8. C. B. Nash, "Psychokinetic control of bacterial growth," *Journal of the American Society for Psychical Research* 51 (1982): 217–331.
9. Braud, *Distant Mental Influence.*
10. Russell Targ, *Limitless Mind* (Novato, CA: New World Library, 2004).
11. PEAR has just recently closed down after decades of research.
12. B. J. Dunne, "Gender differences in human/machine anomalies," *Journal of Scientific Exploration* 12, 1 (1998): 3–55.
13. Dean Radin, "Exploring relationships between random physical events and mass human attention: Asking for whom the bell tolls," *Journal of Scientific Exploration* 16, 4 (2002): 533–48.
14. Dean Radin, *The Conscious Universe: The Scientific Truth of Psychic Phenomena* (New York: HarperCollins, 1997).
15. Ibid., 177–78.

16. J. M. Schwartz, P. W. Stoessel, L. R. Baxter, et al., "Systematic changes in cerebral glucose metabolic rate after successful behavior modification treatment of obsessive-compulsive disorder," *Archives of General Psychiatry* 53 (1996): 109–13.
17. This work was presented in February 2005 at the annual meeting of the American Association of the Advancement of Science.

Chapter 7: Was She Out of Her Mind or Just Out of Her Body?

1. Edward Kelly, Emily Kelly, et al., *Irreducible Mind* (Lanham, MD: Rowman and Littlefield, 2007).
2. C. S. Alvarado, "Out-of-body experiences," in Etzel Cardeña, Steven J. Lynn, and Stanley Krippner (eds.), *Varieties of Anomalous Experience* (Washington, DC: American Psychological Association, 2000), 183–218.
3. C. S. Alvarado, "ESP and out-of-body experiences: A review of spontaneous studies," *Parapsychology Review* 14, 4 (1983): 11–13.
4. Alvarado, "Out-of-body experiences."
5. H. Ehrsson, "The experimental induction of out-of-body experiences," *Science* 317 (2007): 1048.
6. Henrik Ehrsson and colleagues did a brain imaging study of someone during the rubber-hand illusion. This illusion happens when a rubber hand is placed to look as though it is extending from one's arm while one's real hand and its connection to the arm are hidden. When both the rubber and real hands are stroked in corresponding locations simultaneously, one experiences the stroking as only occurring in the location of the rubber hand. Test subjects also flinched when Ehrsson threatened to smash the rubber hand with his fist and were surprised when

they couldn't move the rubber hand's fingers. Even though the subjects knew about the setup, they still had those reactions to the sensory illusion. Brain imaging showed increased activity in the premotor cortex, which processes sensation such as pain, touch, vibration, proprioception (the sense of where a body part is in space), and temperature. The information from the visual system not only influenced the premotor cortex, it was given more weight in deciding where one's hand was.

7. R. L. Morris, S. Harary, et al., "Studies of communication during out-of-body experiences," *Journal of the American Society for Psychical Research* 72 (1978): 1–21.
8. C. Tart, "A psychophysiological study of out-of-the-body experiences in a selected subject," *Journal of the American Society for Psychical Research* 62, 1 (1968): 3–27.
9. M. A. Persinger, W. G. Roll, et al., "Remote viewing with the artist Ingo Swann: Neurological profile, electroencephalographic correlates, magnetic resonance imaging (MRI), and possible mechanisms," *Perceptual and Motor Skills* 94 (2002): 927–49.
10. The activity was described as "7 hertz spike and slow wave."
11. O. Blanke, S. Ortigue, T. Landis, et al., "Stimulating own-body perceptions," *Nature* 419 (2002): 269–70.
12. The right angular gyrus is associated with visual-spatial processing and the orientation of the body in space. The right angular gyrus is almost always significantly smaller than the left angular gyrus, which is associated with language.
13. O. Blanke T. Landis, et al., "Out-of-body experiences and autoscopy of neurological origin," *Brain* 12, 2 (2004): 243–58.
14. M. Niznikiewicz et al., "Abnormal angular gyrus asymmetry in schizophrenia," *American Journal of Psychiatry* 157 (2000): 428–37.

15. S. J. Blackmore, "Out-of-body experiences in schizophrenia: A questionnaire survey," *Journal of Nervous and Mental Disease* 174 (1986): 615–19.
16. Stephen LaBerge, *Lucid Dreaming* (Los Angeles: Tarcher, 1985).
17. K. Nelson, M. Mattingly, and F. Schmitt, "Out-of-body-experience and arousal," *Neurology* 68, 10 (2007): 794–95.
18. These hallucinations and sleep paralysis are two of the four symptoms of narcolepsy, but they can also occur in people whose sleep cycle has been seriously disturbed, such as after a long bout of sleep deprivation. Epidemiological studies of the prevalence of sleep paralysis in the general population have yielded rates of 6 to 11 percent.
19. Phyllis M. H. Atwater and David H. Morgan, *The Complete Idiot's Guide to Near-Death Experiences* (Indianapolis, IN: Alpha Books, 2000).
20. Kenneth Ring, *Heading Toward Omega: In Search of the Meaning of the Near-Death Experience* (New York: Morrow, 1984).
21. J. E. Owens, E. W. Cook (Kelly), and I. Stevenson, "Features of 'near-death experience' in relation to whether or not patients were near death," *Lancet* 336 (1990): 1175–77.
22. K. Ring, and S. Cooper "Near-death and out-of-body experiences in the blind: A study of apparent eyeless vision," *Journal of Near-Death Studies* 16 (1997): 101–47.
23. J. E. Whinnery, "Psychophysiologic correlates of unconsciousness and near-death experiences," *Journal of Near-Death Studies* 15 (1997): 231–58.
24. Atwater and Morgan, *The Complete Idiot's Guide to Near-Death Experiences.*
25. Michael Sabom, *Recollections of Medical Death: A Medical Investigation* (New York: Harper and Row, 1982).

26. Michael Sabom, *Light and Death: One Doctor's Fascinating Account of Near-Death Experiences* (Grand Rapids, MI: Zondervan, 1998).
27. M. A. Persinger, "Near-death experiences: Determining the neuroanatomical pathways by experiential patterns and simulation in experimental settings," in L. Besette (ed.), *Healing: Beyond Suffering or Death* (Chabanel, Quebec: MNH, 1994), 277–86.
28. Glutamate is an excitatory neurotransmitter, which means it increases brain activity. Although ketamine blocks glutamate receptors, this does not result in decreased activity throughout the brain. It decreases activity in the frontal lobe but has a secondary effect of increasing the release of glutamate and the activity in the temporal lobes. J. F. Deakin et al., "Glutamate release and the neural basis of the subjective effects of ketamine," *Archives of General Psychiatry* 65, 2 (2008): 154–64.
29. Many of these drugs are extremely toxic and overuse can cause vacuoles, or tiny holes, in the brain. Women are more susceptible than men, perhaps because their limbic systems are slightly different.
30. K. L. R. Jansen, "The ketamine model of the near-death experience: A central role for the N-methyl-D-aspartate receptor," *Journal of Near-Death Studies* 16 (1997): 5–26.
31. J. C. Gillin, J. Kaplan, et al., "The psychedelic model of schizophrenia: The case of N,N,-dimethyltryptamine," *American Journal of Psychiatry* 133 (1976): 203–8.
32. Richard Strassman, *DMT: The Spirit Molecule* (Rochester, VT: Park Street Press, 2001).
33. The pineal gland contains the highest levels in the brain of the ingredients necessary to make DMT. These include the highest levels of serotonin anywhere in the body, the enzymes

to convert serotonin to tryptamine, and extremely high concentrations of the enzymes (methyltransferases) essential for converting tryptamine to DMT.

34. J. C. Callaway, H. Morimoto, J. Gynther, et al., "Synthesis of (3H) pinoline, an endogenous tetrahydro-beta-carboline," *Journal of Labelled Compounds and Radiopharmaceuticals* 31 (1992): 355–64.

35. The major biological clock in our body is the suprachiasmatic nucleus (SCN), which is a cluster of cells in the limbic system's hypothalamus. The SCN receives information about light from the retinas at the back of the eyes along a bundle of connecting fibers called the retinohypothalamic tract. The SCN then sends signals back to the pineal that influence the production of melatonin.

    The retinohypothalamic tract cannot be the only way that light influences the pineal. A study on blind mice with no retinal receptors (receptors in the back of the eye) found that their production of melatonin was normal. This suggested that the pineal may be able to directly sense sunlight. Other evidence for this comes from light therapy, which is used to treat seasonal affective disorder, a form of depression that occurs as a result of shorter days during fall and winter months. The light effective for treatment needs to be of the same lux (intensity) as sunlight, but it can be shined at the back of one's knees, instead of the eyes, and still have the same beneficial effect. Light therapy's benefits are believed to be a result of resetting the biological clock.

36. The term *energy/matter* is used because energy and matter can be converted into each other.

37. Atwater and Morgan, *The Complete Idiot's Guide to Near-Death Experiences.*

38. J. Doyon and B. Milner, "Right temporal lobe contribution to global visual processing," *Neuropsychologia* 29 (1991): 343–60.

Chapter 8: Evolution and Extraordinary Human Abilities

1. René Warcollier published a book titled *La Télépathie* in 1921 and another titled *Mind to Mind* in 1948. He believed that the errors found commonly in telepathic picture drawings were a clue to telepathy's mechanism. The pictures' components were rearranged into patterns that struck him as similar to the way our dreams process information. They revealed a nonlinear and more primary process than ordinarily thought.
2. Rupert Sheldrake, *Dogs That Know When Their Owners Are Coming Home* (New York: Three Rivers Press, 1999).
3. J. Tolaas, "Vigilance theory and psi. Part I: Ethological and phylogenetic aspects," *Journal of the American Society for Psychical Research* 80, 4 (1986). Although Tolaas's theory was only about mammals, it could apply to birds as well. The theory was first put forth by Montague Ullman in the 1950s.
4. The basis for his statement is unclear, but it may be from the examination of premature babies.
5. Phyllis M. H. Atwater and Dave H. Morgan, *The Complete Idiot's Guide to Near-Death Experiences* (Indianapolis, IN: Alpha Books, 2000).
6. M. Jouvet, "The function of dreaming: A neurophysiologist's point of view," in *Handbook of Psychobiology* (New York: Academic Press, 1975), 499–527.
7. A small percentage of people have language in the right brain. Some have it in both sides of the brain.

8. A. R. Braun et al., "Regional cerebral blood flow throughout the sleep-wake cycle. An H_2O PET study" *Brain*, 120 (1997): 1173–97.
9. G. Baylor, "What do we really know about Mendeleev's dream of the periodic table? A note on dreams of scientific problem solving," *Dreaming* 11, 2 (2001): 89–92.
10. O. T. Benfrey, "August Kekulé and the birth of the structural theory of organic chemistry in 1858," *Journal of Chemical Education,* 35 (1958), 21–23.
11. Daniel Tammet, *Born on a Blue Day* (New York: Free Press, 2006).
12. Pi is the mathematical constant that provides the circumference of circles when it is multiplied by their diameter. It is usually rounded up to 3.14, but the numbers after the decimal point appear to go on indefinitely.
13. Peek had read nine thousand books by 2006, when he was fifty-five years old. The number reported in 2007 was twelve thousand.
14. B. Rimland and D. Fein, "Special talents of autistic savants," in L. K. Obler and D. Fein (eds.), *The Exceptional Brain* (New York: Guilford Press, 1988), 474–92. Darold A. Treffert, *Extraordinary People* (London: Bantam, 1989).
15. T. L. Brink, a psychologist at Crafton Hills College in California, published a study in 1980 that suggested that changes to the left hemisphere were important to the savant syndrome. A normal nine-year-old boy became mute, deaf, and paralyzed on the right side when a bullet damaged his left brain. He developed savant mechanical skills after the accident and was able to repair multigeared bicycles and to invent various contraptions.
16. H. Koshino, P. Carpenter, et al., "Functional connectivity in an

fMRI working memory task in high-functioning autism," *Neuroimage* 24 (2005): 810–21.

17. Synesthesia has been intriguing to scientists for three centuries. Famous synesthetes include the Russian author Vladimir Nabokov, German painter Wassily Kandinsky, American artist David Hockney, and Russian composers Alexander Scriabin and Nikolaẏ Rimsky-Korsakov. People familiar with their work can often see the way in which synesthesia has shaped it. For example, Nabokov frequently uses mixed senses in his descriptions of things. I first read about synesthesia in Cytowic's *The Man Who Tasted Shapes*, which was about a man who consistently felt specific geometric shapes on his tongue in association with specific tastes. For example, whenever he tasted mint, he also felt cool glass columns. That form of synesthesia is one of the rarest. The most common (121 out of 165 studied cases) is seeing colors surrounding letters, which is called lexical synesthesia. For example, one synesthete reports that the letter *b* is navy blue, *a* is gray-blue, and *c* is tawny crimson.
18. V. S. Ramachandran and E. M. Hubbard, "Hearing colors, tasting shapes," Scientific American.com, April 13, 2003.
19. Aleksandr Luria, *The Mind of a Mnemonist* (Cambridge, MA: Harvard University Press, 1968).
20. More recent studies used positron emission tomography, which also uses radioactivity to identify the relative activity of brain regions. These studies were of a more common form of synesthesia (colors linked to words) and demonstrated dramatic increases in both the visual and language processing cortices. Baron-Cohen noted that infants have a similar dramatic increase in cortical activity in response to noise and visual stimulation as the synesthetes. He suggested that we are all born

synesthetes, but that our sensory systems become segregated after the age of six months. He suggested that the bridges between these areas remain in synesthetes. This agrees with Donna Eden, author of *Energy Medicine*, who believes that we are all born seeing auras, but this ability disappears for most of us in our first years. S. Baron-Cohen, J. Harrison, J. Goldstein, and M. Wyke, "Coloured speech perception: Is synaesthesia what happens when modularity breaks down?" *Perception* 22 (1993): 419–26. Donna Eden with David Feinstein, *Energy Medicine* (New York: Penguin Putnam, 1998). The comment about auras was made during Eden's presentation at the Science and Consciousness conference in Albuquerque in April 2004.

21. Grossenbacher was quoted by Brad Lemley in his article "Do you see what they see?" *Discover*, December 1999, 87.
22. J. P. McKane and A. M. Hughes, "Synaesthesia and major affective disorder," *Acta Psychiatrica Scandinavica* 77, 4 (1988): 493–94.

Chapter 9: The Compartmentalization of Consciousness

1. The terms *right* and *left cerebral hemisphere* (as in the quote by Sperry that begins this chapter) are synonymous with *right* and *left brain*. Both are composed of the cortex of the frontal, temporal, parietal, and occipital lobes. The cortex is the outermost section of the brain and is associated with the highest level of processing information.
2. Although the left hemisphere is associated with language for most people, this is only true for 90 to 95 percent of right-handed adults and 60 percent of left-handers. Of the remaining left-handers, half have language in their right hemisphere

and half have it divided between the hemispheres. Also, the tendency to have language predominantly on one side is more the case for heterosexual, nonmusical, right-handed men. Women, homosexual men, and musicians have language divided between the hemispheres, even if it is still more heavily represented on one side. This presence of language in the right hemisphere may be why those three groups, in general, are more in touch with their emotions and have a slightly higher expressed interest in spirituality than nonmusical, heterosexual, right-handed men.

3. Functions were initially assigned by looking at the results of specific damage. Also the brain waves recorded by EEG show differences depending upon which half is more active. The more active hemisphere will usually have more beta waves, and the less active will have more alpha waves. More recent brain imaging technology has confirmed these distinct differences.

4. Decades ago patients' right and left hemispheres were disconnected surgically from each other to prevent the spread of seizures. Superficially the patients appeared and felt normal. Because the brain's internal communication was compromised, each side of the brain could be tested independently of the other. When the right hand and eye were allowed to examine an object, they could name it but not explain what it was used for. The right side of the body uses the left brain, which has the capacity to use language but not understand an object's function. The left hand and eye use the right brain and could demonstrate the object's use, but not name it because the right brain understands function and not language. Those of us with intact connections between the hemispheres name an object held by our left hand by retrieving the name from the other side. The "split brain" patients also had two self-images.

One patient with some language capacity in the right brain was asked to point to the name of his preferred career. His left brain (right hand) pointed to "draftsman" and his right brain (left hand) pointed to "racing car driver." His left brain picked a pragmatic choice and the right picked his dream. Some of the "split brain" patients had one hand ignorant of what the other hand was doing. One patient tried to take his pants off with one hand while he was putting them on with the other. Another man grabbed his wife and shook her violently with his left hand while his right hand tried to intervene. A third patient smacked her dog with her left hand while petting it with her right. The left hand is controlled by the right hemisphere, which is the more aggressive or violent hemisphere. The origin of the word for left, *sinistral*, has the same roots as the word *sinister* and may derive from ancient observations of people with brain damage who had a more violent left hand. Controlling the more aggressive right brain might be one reason the left brain dominates over the right and can keep the right brain's intentions unconscious. Rosadini did experiments with normal subjects in which he alternately placed each brain half under anesthesia by injecting sodium amytal into the left or right carotid artery. When only the right brain was awake, the subject was more depressed. The subjects were euphoric when only the left brain was awake. Dimond's research showed that the right brain tended to regard films as more "unpleasant" or "horrific" than the left. The left frontal lobe is more responsive to joyful music, and the right frontal lobe responds more to fearful and sad music. Research on newborns by Richard Davidson has shown that these emotional differences between the brain halves are present at birth and not learned. This difference can be demonstrated as the case for everyone by the following: First take a picture of someone's face. Then create a picture that is the mirror image of it. Cut both pictures in half and put the two left halves together and the two right halves together. The left side of the face is controlled by the right brain and displays more intense emotions, so the composite of the two left halves looks more emotional than the composite of the right halves, especially if the person was feeling negative emotions. One reason the Mona Lisa's smile is called "enigmatic" is because she is smiling only on the left side of the face, which is controlled by the right brain. We subconsciously expect the happier side of the face to be on the right. S. J. Dimond, L. Farrington, P. Johnson, "Differing emotional response from right and left hemisphere," *Nature* 261 (1976): 690–92; R. J. Davidson and N. A. Fox, "Asymmetrical brain activity discriminates between positive and negative affective stimuli in human infants," *Science* 218 (1982): 1235–37; G. Rosadini and G. F. Rossi, "On the suggested cerebral dominance for consciousness," *Brain* 90 (1967): 101–12.

5. The active use of the right brain in Hebrew may be a contributing factor to its sacredness and powerful role in religious study and rituals, since the right brain is involved in spirituality.
6. This is speculation, but the differences in the direction that phonetic languages are written is probably a result of how the brain is wired. Since the right brain is needed when there are no vowels, people probably found it easier to read when they wrote it this way, just as most people tend to hold the phone to their left ear, which activates the left brain where language is stored for most people.
7. Left brain damage is much more common than right brain damage because of differences in circulation between the two sides. At least four out of five strokes occur on the left, thereby

affecting the right side of the body. The left brain is also more susceptible in the womb and during traumatic births.

8. In general people with right brain damage can't pick the correct punch line for jokes or detect sarcasm. If the phone connection is poor and they miss one out of five words, they will often have a harder time following the conversation than someone with an intact right brain. Right brain damage can also cause severe neglect of the left half of their world. For example, if asked to draw a clock, they will place all twelve of the numbers on the clock's right side. They may even deny that the left half of their body belongs to them. There is a famous case of a man who kept throwing himself out of bed after a right brain stroke because he thought that his left side was another man in bed with him. Right brain damage can also cause one to lose the ability to organize one's perceptual world. Jürgen Lange studied such patients and found that they were rendered totally dysfunctional. They couldn't determine if objects were in the foreground or background. They were so disoriented that they were unable to walk, stand, or even say where they were. Robert Ornstein, *The Right Mind: Making Sense of the Hemispheres* (New York: Harcourt Brace & Company, 1997), 59.
9. The brain waves during meditation were predominantly in the range of delta, 1 to 3 cycles per second (cps), and theta, 4 to 7 cps, instead of the usual alpha range of 8 to 12 cps or beta range of 13 to 30 cps.
10. Mihaly Csikszentmihalyi, *Flow: The Psychology of Optimal Experience* (New York: Harper and Row, 1990).
11. E. G. d'Aquili and A. Newberg, "The neuropsychology of aesthetic, spiritual & mystical states," in R. Joseph (ed.), *Neurotheology: Brain, Science, Spirituality, Religious Experience* (San Jose, CA: University Press, 2002), 243–50.

12. J. Schlag, M. Schlag-Rey, C. Peck, et al., "Visual responses of thalamic neurons depending on the direction of gaze and the position of targets in space," *Experimental Brain Research* 40 (1980): 170–84.
13. James Austin, *Zen and the Brain* (Cambridge, MA: MIT Press, 1999), 491.
14. T. Ono, K. Nakamura, M. Fukuda, et al., "Place recognition responses of neurons in monkey hippocampus," *Neuroscience Letters* 121 (1991): 194–98.
15. The amygdala is the part of the limbic system involved in latent inhibition.
16. Change blindness was first explored by George McConkie in the late 1970s, but the term was introduced by Ronald Rensink in 1997. George W. McConkie and Christopher Currie, "Visual stability across saccades while viewing complex pictures," *Journal of Experimental Psychology: Human Perception and Performance* 22, 3 (1996): 563–81; Ronald Rensink, Kevin O'Regan, and James Clark, "To see or not to see: The need for attention to perceive changes in scenes," *Psychological Science* 8, 5 (1997): 68–73.

Chapter 10: Consciousness and the Web of Life

1. C. G. Jung, "Synchronicity: An acausal connecting principle," in *The Collected Works of C. G. Jung* (Princeton: Bollingen, 1973), 8:15.
2. There are "introvertive" and "extrovertive" mystical experiences. In the former, all physical and mental objects seem to disappear and a unified consciousness emerges. It is accompanied by feelings of bliss and peace. In the latter, material objects

do not appear to vanish but are perceived as having an underlying unity.

3. Thomas Cleary, *Entry Into the Realm of Reality: A Guide* (Boston: Shambhala Publications, 1989), 1190–91.
4. Newberg and colleagues used SPECT (single-photon emission computed tomography) to measure the brain activity.
5. Our brains usually create a singular perspective of the world and don't allow us to see the interconnections between objects. Changes in brain activity can fragment our singular perspective. The fish analogy reminded me of my patient's OBEs in which he witnessed scenes from both his usual in-body perspective and from up above.
6. Fritjof Capra, *The Tao of Physics*, fourth edition (Boston: Shambhala Publications, 2000), 68.
7. There are certain necessary parameters. The synchronization between two pendulums will stop if they are placed more than six feet apart or turned at 90-degree angles from one another.
8. An object will preferentially pick out its resonant frequency when stimulated by a complex combination of frequencies. It will vibrate at its own frequency and filter out the other frequencies. If you strike a hanging spring with a brick attached to the bottom, it will initially bob around erratically but eventually settle down into a rhythm that is its natural frequency. Vibrations at a resonance frequency can have very powerful effects. They can create a feedback loop that magnifies the signal. This is why glass shatters when a sound wave reaches it at its resonance frequency. Nikola Tesla, the inventor of AC current and AC generators, almost brought down a building by applying vibration to it at its resonance frequency.
9. Atoms are broken down into electrons, neutrons, and protons; and the neutrons and protons are broken down into

quarks. The level below quarks is comprised of "strings," according to string theory.

10. At temperatures close to absolute zero, matter is capable of becoming a superconductor, because there is no resistance. The resultant matter is not a gas, liquid, solid, or plasma. The atoms merge into a blurry blob without separate identities. This blob is called a Bose-Einstein condensate. Predicted by Satyendra Bose and Albert Einstein in 1925, it was proved in 1995 when Eric Cornell and Carl Wieman were able to chill matter down to 20 billionths of a degree above absolute zero.
11. The simplest experimental evidence for the existence of zero-point energy in quantum field theory is the Casimir effect. This effect was proposed in 1948 by Dutch physicist Hendrik B. G. Casimir. Feynman is quoted in Lynne McTaggart's *The Field: The Quest for the Secret Force of the Universe*, (New York: Harper Collins, 2002), 23–24.
12. For example, any two points can become connected by drawing a line through them, which increases the dimensions from zero (a point has zero dimensions) to one (a line has one dimension). We can't see the connecting line if our vision is limited to zero dimensions; we would only see points. Three points can be connected in space to form a triangle, which increases the dimensions from one to two. But if we can't see that second dimension, we only see a line and a point. A fourth dot above the triangle can be connected to form a pyramid, but we won't see it if we can't see the third dimension.
13. Much of what we know about the Big Bang comes from mathematical calculations using the current expansion rate of the universe, measurements of the universe's background radiation, and the way in which elements are distributed in the universe. Scientists have concluded from these that 96 percent

of the universe's matter and energy are not perceptible to us by any direct means, which is why it is called "dark."

14. Aspect became the recipient of the 2005 CNRS (Centre National de la Recherche Scientifique) gold medal for his work on entanglement.
15. Zhi Zhao et al., "Experimental demonstration of the five-photon entanglement and open-destination teleportation," *Nature* 430 (2004): 54–58.
16. The headlines jokingly referred to the potential of teleportation as "Beam me up, Scotty." We are a long way from teleportation, but teleportation would work as follows: If two particles, A and B, are brought close enough together that they become entangled, they will function as one unit even when they are moved far apart. When a third particle, C, transfers information to A, C's characteristics automatically become transferred to B as well.
17. Douglas Hofstadter is the physicist who won the 1980 Pulitzer Prize for his nonfiction book *Godel, Escher, Bach: An Eternal Golden Braid.*
18. It is also known as "sensitive dependence on initial conditions."
19. This retention of the same shape is referred to as "self-similarity." It has been popularized by computer-generated, self-similar images known as "fractals." Fractal structures are prevalent in nature. The logarithmic spiral shape of a pine cone is one of many natural expressions of a mathematical sequence called the Fibonacci sequence. This mathematical sequence is derived from adding the previous two numbers of the sequence to one another to create the next, and so on down the sequence. The first nine numbers of the sequence are 1, 1, 2, 3, 5, 8, 13, 21, and 34. The fact that each number is derived from what preceded it creates the self-similarity.

Although this addition of two adjacent numbers is a very simple version of self-reference, it still creates numbers that don't appear ordered until they are mapped out. This is in contrast to 1, 2, 3, 4, 5, 6, etc., which appears ordered without graphing because each number is undergoing the same degree of change regardless of where it is in the sequence.

20. Scientists were mystified by these accounts of fireflies and published twenty articles on this phenomenon in the journal *Science* between 1915 and 1935.

Chapter 11: The Essence of Time

1. *The Autobiography of Charles Darwin*, Francis Darwin, ed. (New York: Dover, 1958), 8–9.
2. Charles Lindbergh, *The Spirit of St. Louis* (New York: Scribner's 1995), 389–91.
3. American tribes with four seasons, such as those in the Plains, often think of time in terms of four stages, with each stage being a preparation for the next.
4. Cited by Tim Folger in "In no time: Searching for the essence of time leads to a confounding question: Does it even exist?" *Discover*, June 12, 2007. The friend was Michele Besso.
5. The most famous of Einstein's *gedanken* experiments was one in which he conceptualized an identical twin traveling in a spaceship at the speed of light while the other twin remained on Earth. When the spaceship twin returned, the twin saw that he had aged significantly less than his earthbound twin. Traveling at the speed of light slows down anything that measures time. It doesn't matter whether the time-measuring clocks are wristwatches, atomic clocks, or people's biological clocks. They

would all slow down if they were traveling at the speed of light because the time intervals, or periods between events, become longer as you approach the speed of light. The time between "ticks" and "tocks" becomes longer, which is called "time dilation." Less time would pass according to the spaceship clock, because there would be fewer ticks and tocks compared to the clock on Earth. Time dilation is confusing until you realize that time, like our weight, is just something we measure. It's easy to see that weight isn't a universal constant. We've seen astronauts float around in space on television because our weight changes when the force of gravity does. Calculations predict this relationship between weight and gravity. But calculations also show a relationship between a time-measuring device's speed of travel and the measurement of time. People haven't traveled at the speed of light, so we haven't seen television footage of a twin age less quickly under those conditions.

6. Einstein also said that time is not separate from space. It forms a continuum called "space-time." As with time, space is also mutable. As an observer's speed approaches the speed of light, the lengths of objects appear different than they do to someone watching them from a stationary position.
7. Einstein compared space-time to a large rubber sheet that has our planets and stars placed like bowling balls on it. The heavier the planet is, the more it bends the rubber sheet, or space-time. Any small objects that enter the curved space-time around the planet would roll toward it, which is what we experience as gravity. We experience gravity as fairly constant, because the mass of the earth that curves space-time is constant whenever we are on the planet's surface. The rubber sheet analogy is not perfect because the curvature of space-time occurs all around the earth and not just underneath it.

In fact, there really is no such thing as "underneath" the earth because there is no up and down in space. It is just as accurate to look at a globe with the Southern Hemisphere on top. Our globe has the Northern Hemisphere on top because the Northern Hemisphere is where the majority of mapmakers and explorers have lived.

8. Another prediction of Einstein's was that clocks would run faster when under less gravitational force. In 1976 a rocket carried a hydrogen-maser clock to an altitude of around six thousand miles, where the effect of gravity is weaker than on Earth. Its timekeeping over a two-hour time period was compared to that of an identical clock on Earth. The clock from the higher altitude had run slightly faster, and the data were within a few parts in a hundred thousand of that predicted by Einstein's equation. Time is measured in a hydrodgen-maser clock by the frequency of emissions of microwaves. The clock on Earth emitted microwaves at 1.42 gigahertz, whereas the clock in the rocket was one hertz faster when it was at an altitude of six thousand miles. Although it might not sound like much difference, the hydrogen-maser clock's emissions do not differ by that much on Earth.
9. Imaginary numbers are defined as numbers that contain the square root of −1. In ordinary math it is impossible to find the square root of a negative number, because the square root of a number is the number that can be multiplied by itself in order to get the number it is the square root of. We can multiply two positive numbers together and get a positive number and we can multiply two negative numbers together, which also results in a positive number. But we need to use imaginary numbers in order to have two identical numbers that we can multiply together to get a negative number.

10. Stephen Hawking, *A Brief History of Time* (New York: Bantam Books, 1988).
11. John W. Dunne, *An Experiment with Time* (Charlottesville, VA: Hampton Roads, 2001).

Chapter 12: The Sum of the Parts Is Greater than the Whole

1. In the 1600s scientists were debating whether light was a wave or a particle. Among the proponents of light being a particle was Pierre Gassendi (1592–1655), who influenced Newton's particle theory about light. The major rationale was that light traveled in straight lines, which was a feature of particles and not waves, which spread outward. Christiaan Huygens (1629–1695) was among those who argued that light was a wave. The fact that light propagated like a wave was proven in 1801 by Thomas Young. Later experiments found that wave/particle duality also applies to subatomic particles, such as electrons.
2. Radin set up a low-intensity laser beam and an interferometer to create the wave interference patterns of light. This device was placed in a double-steel-walled, shielded chamber in the laboratory with a digital camera to record the light patterns.
3. D. Radin, "Intention and reality: The ghost in the machine returns," *Shift: At the Frontiers of Consciousness,* June–August 2007, 23–26.
4. John Wheeler proposed a double-slit experiment called "delayed choice" that added an extremely fast shutter in front of one of the slits. It would open or close in front of the slit during that brief period of time after the photon had already gone through the slit but before the light reached the camera.

The resulting light pattern depended upon whether the shutter was blocking the slit at the time the photon reached the camera and not upon whether it blocked the slit when the photon was passing through. This was the opposite of what one would expect. It was as though the photon's response was changed after the fact. It illustrates that our sense of sequential time is incorrect.

5. J. Wheeler, "Is physics legislated by cosmogony?" in C. J. Isham, Roger Penrose, D. W. Sciama (eds.), *Quantum Gravity: An Oxford Symposium* (Oxford: Clarendon, 1975), 538–77.
6. C. George Boeree, Ph.D., is a professor of psychology at Shippensburg University of Pennsylvania, and this quote is from Wikipedia's definition of collective unconscious.
7. The contents of the collective unconscious are called archetypes. Examples of archetypal symbols include the rose, snake, and sun. The rose is frequently associated with romance, the sun with a god, and the snake with rebirth. Archetypal themes include the passage from innocence to experience and the heartbreak of unrequited love. Archetypal characters include the blood brother, rebel, hero, femme fatale, and wise grandparent. Archetypal patterns of action and plot include the quest, the descent to the underworld, and the feud.
8. The Egyptian wall painting that depicts reflexology is dated at about 2330 B.C.
9. In medicine the plural of *iris* is *irides*, as opposed to the plural of the flower, which is *irisis*.
10. From www.lifesci.sussex.ac.uk/home/Alan_Garnham/Teaching/LTM/mem2.ppt/.
11. R. Lewin, "Is your brain really necessary?" *Science* 210 (1980): 1232–34.

12. Gopi Krishna, *Kundalini: The Evolutionary Energy in Man* (London: Shambhala, 1971).
13. Fluctuations in the earth's magnetic field have a negative correlation with psychic results. Psychic results increase threefold over their average value when experiments are done at 13:30 local sidereal time (LST), plus or minus an hour. One proposed interpretation is that the central part of the galaxy is below the horizon at 13:30 LST, which means that any radiation effects from it are shielded by the earth. Another reason this might be a favorable time would be because of our biological clocks, which affect our physiology. For example, there are several hormones, such as cortisol, that vary widely through the daily cycle.
14. From www.gaia.com/quotes/Arthur_Schopenhauer/.
15. Quoted in Howard Eves, *Mathematical Circles Adieu: A Fourth Collection of Mathematical Stories and Anecdotes* (Boston: Prindle, Weber and Schmidt, 1977), 60.

INDEX

NOTE: Italic page numbers indicate illustrations.

A NOTE ON THE AUTHOR

DIANE HENNACY POWELL, M.D., completed her training in medicine, neurology, and psychiatry at Johns Hopkins University School of Medicine. She also trained at Queen Square and the Institute of Psychiatry in London, and was on faculty at Harvard Medical School before leaving academia for private practice. Dr. Powell has been a member of a part-time think tank on consciousness at the Salk Institute in La Jolla, California. Her articles have appeared in neuroscience and neuropsychiatry journals, and "We Are All Savants" was published in *Shift*, the quarterly journal of the Institute of Noetic Sciences (IONS). She was the lead author of a special publication for IONS called *The 2007 Shift Report*. Dr. Powell lives in Medford, Oregon, and Los Angeles, California. Visit www.dianehennacypowell.com.